This book examines the implications of possible climate change and variability on both global and regional water resources. Water is going to be one of the key, if not the most critical, environmental issue in the twenty-first century because of the escalation in socio-economic pressures on the environment in general. Any future climate change or climate variability will only accentuate such pressures.

The volume initially examines the subject from the perspective of the Intergovernmental Panel on Climate Change (IPCC), and infers possible changes in hydrological regimes and water quality based on the outputs from various scenarios of General Circulation Models (GCMs). In subsequent chapters, the possible effects of climate change on the hydrology of each of the continents is examined. The book concludes with an overview of hydrological models for use in the evaluation of the impacts of climate change. This is the first volume to provide a multidisciplinary framework for synthesizing many diverse trains of thought, research and practice in the assessment of the impacts of climate change on water resources. It will provide a valuable guide for environmental planners and policy-makers, and will also be of use to all students and researchers interested in the possible affects of climate change.

JAN C. VAN DAM is Emeritus Professor of the Delft University of Technology, Delft, The Netherlands. He is also a guest lecturer at the International Institute for Infrastructural, Hydraulic and Environmental Engineering, Delft, The Netherlands.

INTERNATIONAL HYDROLOGY SERIES

The **International Hydrological Programme** (IHP) was established by the United Nations Educational, Scientific and Cultural Organisation (UNESCO) in 1975 as the successor to the International Hydrological Decade. The long-term goal of the IHP is to advance our understanding of processes occurring in the water cycle and to integrate this knowledge into water resources management. The IHP is the only UN science and educational programme in the field of water resources, and one of its outputs has been a steady stream of technical and information documents aimed at water specialists and decision-makers.

The **International Hydrology Series** has been developed by the IHP in collaboration with Cambridge University Press as a major collection of research monographs, synthesis volumes and graduate texts on the subject of water. Authoritative and international in scope, the various books within the Series all contribute to the aims of the IHP in improving scientific and technical knowledge of fresh water processes, in providing research know-how and in stimulating the responsible management of water resources.

INTERNATIONAL HYDROLOGY SERIES

Impacts of Climate Change and Climate Variability on Hydrological Regimes

Edited by

Jan C. van Dam

CAMBRIDGE
UNIVERSITY PRESS

PUBLISHED BY THE PRESS SYNDICATE OF THE UNIVERSITY OF CAMBRIDGE
The Pitt Building, Trumpington Street, Cambridge, United Kingdom

CAMBRIDGE UNIVERSITY PRESS
The Edinburgh Building, Cambridge CB2 2RU, UK http://www.cup.cam.ac.uk
40 West 20th Street, New York, NY 10011-4211, USA http://www.cup.org
10 Stamford Road, Oakleigh, Melbourne 3166, Australia

First published 1999

Printed in the United Kingdom at the University Press, Cambridge

Typeset in Times $9\frac{1}{2}$/13pt, in 3B2 [KW]

A catalogue record for this book is available from the British Library

Library of Congress Cataloguing in Publication data

Impacts of climate change and climate variability on hydrological regimes / edited by Jan C. van Dam
    p. cm. – (International hydrology series)
  Includes bibliograpical references and index.
  ISBN 0-521-63332-X
  1. Hydrology – Enviromental aspects. 2. Climate changes – Enviromental aspects. I. Dam, Jan C. van, 1931.
 II. Series.
 GB661.2 I45 1999
  551.48 – dc21 98-25768 CIP

ISBN 0 521 63332 X hardback

Contents

Contributors

Dr Nigel W. Arnell
Department of Geography, University of Southampton, Southampton, UK

Dr Mike Bonell
Division of Water Sciences, UNESCO, Paris, France

Prof. Dr Eng. Benedito Braga, Jr
Department of Hydraulic and Sanitary Engineering, Escola Politécnica da USP, 05508-900 Sao Paulo, SP, Brazil

Prof. Dr iR Jan C. van Dam
Emeritus Professor, Delft University of Technology, Delft, The Netherlands

Valerie Jones

Dr George H. Leavesley
Water Resources Division, US Geological Survey, Denver, Colorado, USA

Dr Thierry Lebel
ORSTOM, Laboratoire d'étude des Transferts en Hydrologie et Environnement (LTHE), Grenoble, France

Prof. Dr Luis C.B. Molion
Department of Meteorology, Universidade Federal de Alagoas, Cidade Universitária, 57072-970 Maceió, Alagoas, Brazil

Dr A.I. Shiklomanov
Arctic and Antarctic Research Institute, Russian Federal Service for Hydrometeorology and Environmental Monitoring, St Petersburg, Russia

Prof. Dr Igor A. Shiklomanov
State Hydrological Institute, St Petersburg, Russia

Dr Jacques Sircoulon
ORSTOM, Paris, France

Executive summary

It is now widely accepted that the increasing concentrations of greenhouse gases in the atmosphere are affecting the Earth's radiation balance, resulting in higher global air temperatures. However, there is still a great deal of uncertainty about the likely effects of such temperature rise on climate, and even more about the impacts of climate change and variability on the world's hydrological regimes and water resources. The interactions between climate change and hydrological conditions will also be affected, either directly or indirectly, by a number of socio-economic factors such as population pressure and changes in land use.

This final report of the Working Group of UNESCO's IHP-IV project H.2.1 provides an overview of ongoing activities in the 'trans-science' domain that involves climatologists, meteorologists, hydrologists, and water resource engineers. In recent years, a number of models of the world's climate have been developed to generate alternative scenarios of future climate conditions. The results of these scenarios are then used by hydrologists to assess the probable effects of climate change on hydrological regimes. Such information is in turn required by policy makers and water resources managers to enable them to formulate appropriate adaptive strategies.

This report is relevant to the international community of scientists and practitioners of hydrology and climatology. It gives an overview of the climatic and hydrological changes that are already occurring around the world, and surveys the variety of methods that are being used to forecast future changes. Such a survey can of course never be complete, but it summarizes the areas for which information was available and accessible.

The preparation of this report partly coincided with that of the second IPCC report, and some members of the Working Group and external assessors were involved in both processes. The two reports are quite different in nature, however. This report is restricted to the hydrological aspects of climate change and their consequences for water resources, and deals in particular with the various methods of 'trans-science' (hydrology and climate) research and their inherent problems. It therefore complements the IPCC report; there is little overlap and there are no major contradictions.

The report is structured as follows. After the Introduction, Chapter 2 provides an overview of the work of the IPCC in the period 1988–94. Chapters 3–7 review the work being undertaken in South America, North America, Europe, Africa, and Asia and Australia, respectively. Chapter 8 is unique in that it provides a concise overview of the models that are being used in the evaluation of the impacts of climate change on hydrology. Finally, Chapter 9 presents the conclusions and recommendations of the Working Group.

CONTINENTAL ANALYSES

Chapters 3–7 detail the results of the various modelling studies that have been undertaken to assess the possible impacts of climate change and variability on hydrological regimes. These continental analyses highlight a number of issues arising from the use of scenarios based on the results of general circulation models (GCMs), including the problems of downscaling the GCM results to provide the inputs for hydrological models at the catchment level, to produce the information required by water resources policy makers, planners and engineers.

South America (Chapter 3): The authors note that there have been few GCM-based studies of South America, except for the Uruguay basin. The results of different GCMs show some marked inconsistencies, particularly in terms of changes in precipitation, and their effects on river discharges. It is clear that GCMs have limited ability to describe accurately the convective precipitation process, and soil–water–plant interactions with the atmosphere in a tropical environment. However, a large-scale field experiment in the Amazon basin (LBA) is now in progress to improve the surface parameterization of GCMs to assess the impacts of deforestation on the Amazonian climate.

North America (Chapter 4): Many regional studies of climate–hydrology relations have provided insights into the potential sensitivity of water resources to climate change, and have also highlighted a number of shortcomings in the process of modelling in general. Such is the degree of uncertainty in the results of both hypothetical and GCM-based scenarios, that they could all be regarded as investigations of the sensitivity of water resources to changes in temperature and precipitation. More robust models are needed that can simulate atmospheric and hydrological processes over a broad range of conditions and spatial and temporal scales. The main obstacles to the development of such models include the limited understanding of many of the hydrological-atmospheric processes involved, and their interactions, and the lack of data with which to develop and test the models of such processes. For example, the temporal and spatial variations in seasonal precipitation, temperature and runoff in parts of North America have been shown to be related to large-scale atmospheric circulation patterns such as the El Niño–Southern Oscillation, which are still only poorly understood. There is also little consistency in the applied methodologies and assumptions. Resource managers will therefore have to continue to rely on scientists' 'best estimates' as the basis for their long-term plans, and must await improvements in the models and methodologies before more reliable estimates are available.

Europe (Chapter 5): In Europe, the seasonal distribution of runoff is more likely to be affected by climate change than the annual totals, especially in snowmelt-dominated catchments. In the milder regions of western Europe, climate change scenarios also indicate that the seasonality of flows will increase, with more frequent floods in winter and more persistent low flows in summer. In southern Europe, some studies have also investigated the possible effects of climate change on groundwater recharge rates. However, there have been few studies of the linkages between changes in land use and hydrological regimes, which may be more important than the impacts of climate change. In response to the uncertainties surrounding climate change scenarios, high-quality runoff data sets for Europe are now being compiled under the FRIEND programme in order to detect hydrological variability.

Africa (Chapter 6): Africa has a wide variety of hydrological regimes, but the Sahel is the most vulnerable to climate change. In both the Sahel and the Nile basin, changes in precipitation will be more important than changes in temperature: less rainfall will lead to less vegetation, leading to overgrazing, leading to less evaporation. The mechanisms of this downward spiral are still inadequately included in GCMs. The Nile basin contains a wide variety of sub-catchments, but such changes in vegetation and soils are still insufficiently included in the models. Southern Africa is now being studied under the FRIEND programme. Studies of the impacts of climate change in Africa are hampered by inadequate instrumentation, and time series data that are too short. HAPEX-type experiments should run for at least five years, and a catchment approach should be adopted. Analyses of satellite data have indicated the importance of the presence of aerosols in the atmosphere. The HAPEX-Sahel integrated database will be freely accessible in 1997.

Asia and Australia (Chapter 7): In a detailed case study of the Caspian Sea, the authors describe the dramatic rise in the level of the sea in recent years due to both climate change (increased precipitation in the Volga basin), and human factors (reduced water abstraction for irrigation and industry). In the cold and temperate regions of Asia, increased air temperatures would have a greater effect on runoff than changes in precipitation, particularly in cold regions. Long-term hydrometeorological data for the Yenisey basin show that so far, maximum warming has occurred in the southern part of the basin, and minimum warming in the north, on the Arctic Ocean coast. These observations are inconsistent with the results of both GCM and paleoclimate scenarios, all of which have forecast greater warming at higher latitudes. Surprisingly high air temperatures are predicted for the Yenisey basin. In the humid tropics, where runoff regimes depend mainly on precipitation, global warming would increase precipitation and runoff, particularly during the wet season. Runoff is thus much more sensitive to changes in rainfall than to changes in potential evaporation. With a 1–2°C increase in global air temperatures, total water resources are likely to increase; total runoff in Asia could increase by 20–30%, and in Oceania by 10–18%.

OVERALL CONCLUSIONS

The overall conclusions of the Working Group are presented in Chapter 9 under the following headings: (a) the use of paleoclimate scenarios; (b) the use of GCM scenarios; (c) hydrological models and climate change; (d) the uncertainties associated with GCMs and hydrological models; (e) field experiments to improve GCMs; (f) climate change and water resources management; and (g) the outcomes of the IPCC process. The main conclusions are as follows:

- The effects of climate variability and change on the hydrological cycle will be coincident with those of changes in land use, which could be of the same order of magnitude. It is therefore difficult to separate these two sets of effects in analysis and synthesis.

- Despite the considerable efforts that have been undertaken in research and modelling climate change and its impacts in recent years, the results are still highly uncertain, for a number of reasons: (1) the unpredictability of future emissions of greenhouse gases and the concentrations of aerosols; (2) the shortcomings in the capabilities of existing GCMs; and (3) the different levels of scale adopted in GCMs and hydrological models. There are two main sources of 'noise': (i) the noise in the GCM results, which can give rise to variable and even contradictory results; and (ii) the noise in the results of different hydrological methods. After more than 15 years of research, progress in reducing both of these sources of noise has been disappointing. A great deal of work is still needed to achieve adequate coupling between the results of GCMs and hydrological models.

- A number of multidisciplinary field experiments have been designed as a means of improving land surface parameterization from the hydrological perspective. However, most of these experiments have tended to focus on the vertical terrestrial–atmospheric exchanges of energy and water vapour, rather than on details of horizontal water fluxes.

- Hydrological models would be considerably improved if they incorporated a more physically based understanding of hydrological processes and their interactions, and if parameter measurement and estimation techniques were developed for applications over a range of spatial and temporal scales.

- Large-scale basin studies such as HAPEX, GCIP and LBA, which incorporate a nested drainage basin approach (NDBA), appear to be the key to improving our understanding of processes and for modelling the hydrological impacts of climate change.

- The inconsistencies in the model outputs mean that policy makers and water resource managers face equally high uncertainty in their task of formulating appropriate adaptive strategies. In the future, such strategic choices will be further complicated by socio-economic factors and the associated changes in the demand for water. In areas of rapid population growth, particularly in developing countries, the availability of water per capita will decrease, irrespective of climate change.

- In 1988 the IPCC was established to gather together and assess the results of research that has been conducted around the world on various aspects of climate variability and change. The Panel was then to use that information to evaluate the environmental and socio-economic consequences of climate change, to formulate realistic response strategies, and, above all, to publicize their findings to the scientific community, international organizations and decision-makers. The IPCC has fulfilled most of these objectives, and in this respect the process has been beneficial and useful to the international community in general. Nonetheless, the Panel has not been able to recommend one future climate change scenario or methodological approach for all assessments of the likely consequences of climate change. There has been little progress in improving the reliability of forecasts of climate change, nor of the probable impacts on water resources, the environment and society.

RECOMMENDATIONS

The recommendations of the Working Group are summarized in the following; the complete text can be found in Chapter 9. Depending on the message, these recommendations are addressed to individual nations, to the international organizations UNESCO-IHP, WMO, IPCC and international funding agencies, to policy makers, climatologists, hydrologists, modellers of climate and hydrology, water resources engineers and planners.

1. *Coordination*: Climate change scenarios and models should be selected so that the results will be directly comparable. This will require international coordination.

2. *Forecasts of future conditions*: More accurate forecasts of likely increases in greenhouse gas concentrations, and of changes in land use, are needed. Such forecasts require international cooperation.

3. *Adaptive measures*: Water resources engineers, planners and policy makers should be alert to any trends in climate as they become clearer, so that timely adaptive measures can be taken.

4. *Sensitivity analysis*: More accurate determination of changes in climate variables is needed, particularly in cases of high hydrological sensitivity and where the availability of water resources is low in relation to the demand.

5. *Water resources management studies*: UNESCO-IHP should frame a programme of water resource management studies and applications, including as inputs the

results of GCMs and subsequent analyses of data on climate variability and change.

6. *Anticipating extreme events*: Greater attention should be given to studies of the effects of extreme hydrological events such as floods, storms in urban areas, and droughts. For vulnerable locations, it may be useful to consider the transposition of existing data from other locations that are representative of the predicted changes in climate.

7. *Large-scale field experiments and modelling approaches*: In large-scale field experiments, greater attention should be paid to quantifying horizontal water fluxes in order to improve the effectiveness of the nested drainage basin approach (NDBA).

8. *Future large-scale field experiments*: Decisions as to whether new large-scale field experiments should be set up in other regions should be taken after weighing the value of the expected results against the inevitably high costs of such experiments.

9. *Freely accessible databases*: Future large-scale studies should include the compilation of complete and freely accessible databases, as is currently being done within HAPEX-Sahel.

10. *Data collection network*: It is important to have access to long-term data sets compiled both in the past and in the future. The WMO and individual nations should therefore recognize that a global network of climate and hydrological stations needs to be maintained at an adequate density. This is particularly the case in areas where data collection has been discontinued due either to natural disasters or political instability.

11. *Groundwater*: It is recommended that an inventory is made of the studies that have been undertaken so far of the impacts of climate change on groundwater regimes. Only then should areas for further study be selected, and appropriate research proposals formulated. Such an inventory and subsequent research proposals might be undertaken as a UNESCO-IHP project.

12. *Funding*: International funding agencies should focus on areas for which the data are scarce, and where the most serious problems, due to hydrological conditions and/or human impacts, are expected. North Africa, particularly the Sahelian region, should be given high priority. Because the process of climate change is slow, and water resources engineers need to pay particular attention to extremes, ongoing and new field experiments need to be established for periods of decades at least.

M. Bonell
J.C. van Dam
V. Jones

Acknowledgements

A draft of this final report of the Working Group on the UNESCO IHP-IV Project H-2.1 was reviewed and discussed during the last meeting of the Working Group, held in Delft, the Netherlands, 15–17 May 1995. The valuable comments of the three external assessors, Professor S. Kaczmarek (Poland), Dr W.J. Shuttleworth (USA) and Dr E.Z. Stakhiv (USA), greatly improved the report. The members of the Working Group are grateful to the external assessors for their constructive criticism.

Dr Bonell, Programme Specialist of the Division of Water Sciences at UNESCO, was appointed to organize the meetings of the Working Group and to provide secretarial support, which he did with great efficiency. He was always to the point and stimulating during the meetings. He also made substantial contributions to the final versions of the Conclusions, Recommendations and Executive Summary of this report.

Valerie Jones did an excellent job in editing the heterogeneous material of this report. Her familiarity with the subject enabled her to trace internal inconsistencies and, together with the Editor's comments, helped to improve the text. We benefited greatly from her input.

I would like to thank the Director of the International Institute for Infrastructural, Hydraulic and Environmental Engineering (IHE), Delft, the Netherlands, for his generous support for this project. In particular the efforts of Mrs E.L.S. Janssen of this institute, during the process of editing and production of the manuscript, are greatly appreciated.

Finally, I would like to thank my fellow members of the Working Group for their kind cooperation.

J.C. van Dam
Editor

1 Introduction

J. C. VAN DAM

1.1 PROBLEM IDENTIFICATION

The subject of UNESCO's International Hydrological Programme (IHP-IV) project H-2.1, 'Study of the relationship between climate change (and climate variability) and hydrological regimes affecting water balance components', is relatively new, but it is of vital importance for society. Climate change and variability will affect the hydrological cycle, which will in turn affect both the distribution and availability of water resources for domestic use, for food production, and industrial activities, as well as for the production of hydropower. Other hydrology-related aspects include flood control, water quality, erosion, sediment transport and deposition, and ecosystem conservation. Most uses of water are economically important, and thus are related to socio-economic development, as well as to public health and well-being and the environment, which themselves are also inter-related. Thus the concerns over the potential impacts of climate change range from the causes to the ultimate and diverse social consequences, which will vary depending on the location and the human responses. This volume addresses mainly the impacts of climate change on hydrological systems, but occasionally some other aspects are also considered.

The social relevance of this study is well illustrated in the tentative statement formulated at a meeting of the Working Group, as follows:

> The world's population is increasing at an unprecedented rate, and this will have implications for many areas of human activity. According to the UN Conference on Population and Development, held in Cairo in 1994, the global rate of population increase is now about 1.6%, whereas for the continent of Africa the rate is as high as 2.8%. Moreover, the proportion of the world's population living in urban areas is increasing rapidly, particularly in the less developed regions. There is now a consensus among scientists that the populations of arid and semi-arid regions are likely to be the most seriously affected by climate change. It is therefore of paramount importance to improve our understanding of how climates may change in the future, and of the various ways in which such changes will impact the populations of these regions.

Research in several areas of geology indicates that the climate has changed throughout the history of our planet. Information dating back many hundreds of millions of years has been obtained from the Earth's geological archive, consisting of the subsoil of the Earth, glaciers and polar icecaps, and the remains of flora and fauna that serve as indicators of paleoclimatic conditions. In contrast, quantitative records of meteorological and hydrological conditions of the past date back a few centuries at most. In between these two extreme time scales, much information has also been obtained from other sources, such as archeological and historical records of the past few thousand years, pollen analyses, and radiocarbon dating (Issar, 1995). The Historical Archives Climate Project is now being designed as a joint project of the World Meteorological Organization (WMO), UNESCO, the International Council of Scientific Unions (ICSU) and the International Council of Archives (ICA), with the objective of expanding our knowledge of climates in the past.

Over the long term, the climate of the Earth has changed due to a number of natural processes, such as gradual variations in solar output, meteorite impacts and, more important, sudden volcanic eruptions in which solid materials, aerosols and gases (sulphur) are ejected into the atmosphere. Ecosystems have adapted continuously to these natural changes in climate, and flora and fauna have evolved in response to the gradual changes in their physical conditions, or have become extinct. The rate of climate change over time has been a major determining factor in the possibility and type of adaptation, or extinction.

Humans have also been affected by and have adapted to changes in local climates, which have occurred very slowly. Over the past century, however, human activities have begun to affect the global climate. These effects are due not only to the population growth, but also to the application of technologies that have been developed for survival, to raise standards

1

of living, and, regrettably, also for non-peaceful purposes. The resultant changes in climate are relatively recent, but have occurred much more rapidly than those due to natural causes. The scale of the present climate forcing is unprecedented, and is due to emissions of greenhouse gases, deforestation, urbanization, and changes in land use and agricultural practices. The increased concentrations of the so-called greenhouse gases – carbon dioxide (CO_2), methane (CH_4), nitrous oxides (NO_x), ozone (O_3) and chlorofluorocarbons (CFCs) – in the atmosphere is causing the air temperature to rise, as in a greenhouse, and this in turn is having many other effects such as changes in the rates of evapotranspiration and in the amounts and distribution of precipitation. Mankind will certainly respond to these changing conditions by taking adaptive measures, such as changing patterns of land use, although it is difficult to predict what adaptive measures will be chosen, and their socio-economic consequences.

Climate change does not necessarily imply that standards of living will fall. For example, increased CO_2 concentrations in the atmosphere may be beneficial for plant growth, thus enhancing agricultural production in some regions, which will have economic and social consequences. This partly explains the different degrees of willingness of various national governments to take action. The level of development of a particular country is another important factor in attitudes to this matter.

This section has outlined the problems and opportunities presented by climate change and variability, and some of the impacts on the hydrological cycle and consequently on water resources. Real world problems require research, so that the role of hydrologists, in cooperation with climatologists, is to describe the physical (meteorological and hydrological) processes in such a way that not only the past and present conditions can be simulated accurately, but that the future conditions can be predicted in terms of averages and frequency distributions for a number of primary climatic and hydrological variables and the derived parameters that can be used to characterize future changes in water resource systems across the globe.

1.2   THE CLIMATE SYSTEM

The climate system refers to the atmosphere, but there are close interactions with the seas and oceans and with the land surface, in terms of fluxes of energy, water and carbon dioxide. The atmosphere has only a small buffer capacity, but the oceans and, to a lesser extent, the glaciers and icecaps are gigantic buffers for heat and water. Thus there is a considerable time lag between the causes of climate change and their effects. Since natural changes in climate occur only very slowly, so also do their effects. Similarly, the full effects of the increase in the intensity of human activities will take some time to be felt on the human time scale, as will the benefits of any countermeasures that may be taken – such as changes in land use – in response to climate change.

1.2.1 Recent studies of climate change

In recent years awareness has been growing that the rate of climate change due to human activities is accelerating. This new realization is due to climate studies, based on meteorological observations over the past few centuries of increasing numbers and types of parameters. Most recently, the issue of climate change and its effects has been addressed by the Intergovernmental Panel on Climate Change in its first report (IPCC, 1990a), its summary for policymakers (1990b), followed by a supplement (IPCC, 1992), and the second IPCC report (IPCC, 1995).

1.2.2 Hydrological impacts of climate change

Climate change and variability has many effects on the hydrological cycle and thus also on water resources systems. Flood protection or water supply systems, both existing and planned, will be affected. These effects may not necessarily be negative, but they need to be anticipated as far as possible (Kaczmarek *et al.*, 1996), because of the socio-economic consequences of the costs of construction, operation and maintenance of water control systems.

Changes in the hydrological cycle due to climate change, whether natural or human-induced, will also lead to changes in natural vegetation patterns, through desertification, the relative rise in sea level, etc. The human response is likely to be to change existing systems of land use, such as by introducing new crops or new cropping patterns with different water requirements, both in total and in their distribution over time. Any change in the hydrological cycle resulting from human responses to changing water resources systems will add to the interactions between the climate and the hydrological cycle. The impacts of human responses on the hydrological cycle are even more difficult to predict than those of climate change.

The planning and construction of water supply systems, coastal protection structures, dikes, etc., take many years, and the lifetimes of such structures are of the order of a century. It is therefore necessary to design them according to criteria that relate to the future, a century ahead, that is, during and after the present period of climate change.

The assessment of climate change and its likely impacts on the hydrological cycle is extremely complex. Hydrologists have to deal with several time scales, varying from periods of several years in dealing with problems of reservoir operation or groundwater levels, to periods of the order of 5–15 minutes in dealing with urban runoff and the capacity of sewerage systems. Intermediate between these two, floods can occur on time scales of days to weeks. Hydrologists therefore need to be informed by climatologists about annual or monthly values (means and standard deviations) of relevant parameters such as precipitation and evaporation (or the underlying climatic variables), and they also need to have detailed statistics of various events or intervals of short duration.

Climate change and variability are occurring at the global or continental levels, i.e. at areal scales that are much larger than those at which hydrological problems are encountered. The impacts are usually felt over areas larger than individual countries or even international river basins. Studies of the relationship between climate change and variability and hydrological regimes therefore need to be international.

The effects of sea level rise due to climate change on the hydrological cycle may be particularly acutely felt in coastal zones. Apart from the loss of land, the regimes of the downstream reaches of rivers, deltaic areas and estuaries are likely to change, and consequently also the distributions of fresh and saline water, both in surface water and in groundwater.

1.3 STATE OF THE ART OF MODELLING

1.3.1 Integrated models

Modelling the hydrological impacts of climate change involves two issues: climate change and the responses of hydrological systems, or rather their mutual interaction. So far, there is no integrated atmosphere–hydrosphere model that can simulate hydrological phenomena at the basin scale for given scenarios taking into account the feedbacks between climate and hydrology. Whether desirable or not, so far it has not been possible to develop such an 'ideal' integrated model for a number of reasons:

- the spatial scales of existing climate models are too coarse to allow for a proper fit to the areas to which hydrological models are commonly applied; as a consequence,
- representations of interactive processes can not yet be properly included in such models, and even if they could,

- the computer time required would make such models impractical.

Two kinds of models are used: the so-called general circulation models (GCMs) to model climate, and hydrological models.

1.3.2 General circulation models (GCMs)

GCMs are used to obtain descriptions of current atmospheric processes. At present, there are only a handful of such models, because they require considerable effort to build, and they can only be implemented and run on very large capacity computers. The few large research institutions that have developed and applied such models include (in alphabetical order)

- the Canadian Climate Center (CCC);
- the Commonwealth Scientific and Industrial Research Organization (CSIRO), Australia;
- the European Centre for Medium-range Weather Forecasting (ECMWF), Reading, UK;
- the Geophysical Fluid Dynamics Laboratory (GFDL), Princeton, NJ, USA;
- the Goddard Institute of Space Sciences (GISS), USA;
- the Laboratoire du Météorologie Dynamique (LMD), Paris, France;
- the Max Planck Institute (MPI), Germany;
- the Meteorological Research Institute (MRI), Japan;
- the National Center for Atmospheric Research (NCAR), Boulder, CO, USA;
- Oregon State University (OSU), OR, USA;
- the United Kingdom Meteorological Office (UKMO), Hadley Centre, Bracknell, UK.

The areas modelled by GCMs are subdivided into grid cells with horizontal dimensions of the order of 300×300 to 1000×1000 km^2. In the vertical direction, the Earth's atmosphere is subdivided into six to ten or more layers, each several hundred metres thick. The temporal resolution is of the order of 30 minutes to 1 hour.

Recently, regional circulation models (RCMs), with grid cells of the order of 50×50 km^2, have been developed for application to relatively small regions, using inputs from GCMs. Obviously, the resolution of an RCM for a particular region is better than that of a GCM.

The atmospheric processes modelled in GCMs are based on mathematically formulated physical laws. The various GCMs differ not only in their grid sizes and the number of layers, but also in the number of processes and relevant parameter values that can be included. A weak point of most GCMs is their inability to model properly the physics of clouds; in this respect the various GCMs vary considerably, and often produce quite different results.

The presence and effects of aerosols originating from human activities and volcanic eruptions are not yet included in the present GCMs, even though there are strong indications that carbon-containing aerosols can have a great influence on climate (Mitchell *et al.*, 1995). According to ten Brink (1996), 'the direct reflection of sunlight by anthropogenic aerosols in western Europe is larger than the (enhanced) greenhouse effect'. Since 'dust particles' compensate for the greenhouse effect, there is clearly a need for further research on the presence and effects of aerosols.

In GCMs the important elements are the interactions between land and water surfaces, and the fluxes of energy, water and CO_2. The descriptions of these processes, for large areas, are often based on small-scale field data or models that cannot be valid for such large grid sizes. This problem of upscaling is another reason for the differing results of GCMs.

1.3.3 Hydrological models

Hydrological models vary enormously, depending on their objectives, e.g. whether they are used to model surface water or groundwater. These can be physically based models, conceptual models, or 'black box' models, and can be either deterministic or partly stochastic (i.e. with deterministic components). The inputs to hydrological models can be either prescribed data, temporal analogue data, spatial analogue data or GCM output data:

- prescribed data, e.g. present-day precipitation, multiplied by a particular factor (multiplicative method), or the same data with an addition (additive method) (Brandsma, 1995);
- temporal analogue data, i.e. data recorded at a particular location over a period in the past in which the climate was wetter or warmer, and more or less the same as what can be expected during or after climate change in the catchment area under consideration;
- spatial analogue data, i.e. data recorded at other locations, where the present climate could be the future climate of the location under consideration. For example, the weather systems of some locations on the west coast of Europe could shift in a northerly direction (Brandsma, 1995).

1.3.4 Coupling of GCMs and hydrological models

The output from GCMs is still not good enough for the purposes of hydrological modelling, for the reasons explained above. There has been some progress in the modelling of clouds and atmosphere–ocean interactions. Atmosphere–land surface interactions are still important areas of research, again because of problems of scale. The grid cells of GCMs are generally much larger than most of the basins in which the interactions between the atmosphere and hydrological cycle are studied by water balance. This means that there is a problem of regionalization or upscaling of the fluxes obtained from hydrological basin studies for use as inputs to GCMs, and/or of downscaling the results of GCMs to individual basins (Feddes *et al.*, 1989; Feddes, 1995). The problem of scaling has long been recognized, and exists independently of climate change. A better understanding of the interactive processes that result in fluxes is therefore essential for better modelling.

1.3.5 Field experiments

The problem of coupling GCMs and hydrological models is currently being addressed in a number of field experiments on various scales, many of which involve international cooperation. Most of these experiments apply remote sensing techniques, the high resolution of which, and the improved correlation between the relevant information on land surface conditions and fluxes, offer good prospects for resolving the problem of scaling. Some of these experiments are briefly illustrated in the following; more complete descriptions, including the results obtained so far, are given for each continent in the relevant chapters in this volume.

- The Hydrologic-Atmospheric Pilot Experiment – Modélisation de Bilan Hydrique (HAPEX-MOBILHY), conducted during 1986 in southwest France. This first large-scale field experiment (Shuttleworth, 1987) provided simultaneous detailed measurements of weather and surface flux variables over agricultural and forested areas. Aircraft and satellite remote sensing data were used for integration to a larger scale.
- The Hydrologic-Atmospheric Pilot Experiment-Sahel (HAPEX-Sahel and HAPEX-Niger) is a similar large-scale experiment that is being conducted in the semi-arid belt of Africa. Here too, remote sensing techniques have been applied.
- Within the framework of the International Satellite Land Surface Climatology Project (ISLSCP), the First ISLSCP Field Experiment (FIFE) was undertaken in

1987–88 in Kansas, USA (Shuttleworth, 1987). It also included satellite data.

- The Anglo-Brazilian Climate Observation Study (ABRACOS), undertaken jointly by British and Brazilian institutions in 1990–94, comprised several small-scale experiments in Brazil but did not include remote sensing data.

- The Large-scale Biosphere–Atmosphere Experiment in Amazonia (LBA),* to commence in 1997, will use remote sensing techniques to examine the tropical climate of the Amazon basin.

- The Biosphere-Atmosphere Transfers and Ecological Research *In situ* Studies in Amazonia (BATERISTA) can be regarded as a complement to the LBA for calibration and validation of the models developed within the LBA experiment.

- Within the Global Energy and Water Cycle Experiment (GEWEX), GEWEX Continental-scale International Projects (GCIPs) will be conducted in the Mississippi River basin (USA) and in the cold region of the Mackenzie River basin (Canada), the Mackenzie GEWEX Study (MAGS).

- Within the European Programme on Climate and Natural Hazards (EPOCH), there is a European International Project on Climate and Hydrological Interactions between Vegetation, Atmosphere and Land Surfaces (ECHIVAL). In the regional ECHIVAL Field Experiment in Desertification-Threatened Areas (EFEDA), now being conducted in central Spain, remote sensing and field data on soil moisture will be related to obtain average areal values.

In the implementation of the GEWEX Continental-scale International Project (GCIP) and the Large-scale Biosphere–Atmosphere Experiment in Amazonia (LBA), a catchment approach will be adopted. The LBA (LBA, 1996) will incorporate a nested drainage basin approach (NDBA; Bonell and Balek, 1993) in order to address the problem of scaling from the micro-, through the meso-, to the macroscale. The NDBA concept, which is the product of a formal agreement between UNESCO's IHP and IGBPBAHC, is described in more detail by Dunne and Barker (1997).

1.4 RELATED ACTIVITIES

Many international governmental and non-governmental organizations are collaborating in a wide variety of research programmes on or related to climate change. The jargon and the acronyms used by these organizations and their research programmes may make it difficult for outsiders to find their way through this labyrinth. Figure 1.1 gives an overview of these programmes and how they are interrelated. Explanations of the many acronyms used in this field are given in the Appendix.

The most important activities are those of the IPCC, as noted in Section 1.2.1. The WMO has undertaken a number of relevant initiatives (Lemmelä *et al.*, 1990; WMO, 1985a, 1989a,b, 1990), and in 1980 launched the World Climate Research Programme (WCRP), with WCRP–Water as one of its sub-programmes. Within UNESCO's IHP-IV programme, a related project (H-1.1) is a 'Review of the scientific aspects of the interface processes of water transport through the atmosphere–vegetation–soil system at an elementary catchment and grid size scale'.

In 1987 the International Association of Hydrological Sciences (IAHS) held a symposium on the subject (Solomon *et al.*, 1987), and set up an International Committee on Atmosphere, Soil and Vegetation Relations (ICASVR, 1992), which operates in cooperation with the H-1.1 project of UNESCO's IHP-IV programme and the International Geosphere–Biosphere Programme (IGBP), a joint programme of the ICSU and the WCRP of the WMO (IHP-ICASVR-IGBP, 1995).

One of the core projects of the IGBP in cooperation with the WCRP is the Biospheric Aspects of the Hydrologic Cycle (BAHC). The activities undertaken within this project, in

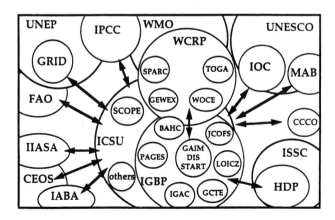

Figure 1.1 Overview of international organizations and their research programmes (modified from Zwerver *et al.*, 1995).

* The recently renamed LBA incorporates the formerly independent experiments LAMBADA (Large-scale Atmospheric Moisture Balance of Amazonia using Data Assimilation), BATERISTA (Biosphere-Atmosphere Transfers and Ecological Research In Situ Studies in Amazonia), and AMBIACE (Amazon Ecology and Atmospheric Chemistry Experiment).

particular those concerned with the relationships between Soil, Vegetation and Atmosphere (SVAT), are relevant to the ongoing projects described in this volume.

Among the multitude of activities and relevant publications, the following are of interest:

- the proceedings of the International Climate Change Research Conference, held in Maastricht, the Netherlands, in 1994 (Zwerver et al., 1995);
- evaporation modelling (Tallaksen and Hassel, 1992);
- an assessment of the impacts of climate change on tropical rainforests (Hulme and Viner, 1995); and
- possible impacts of climate variability and change on tropical forest hydrology (Bonell, 1998).

In particular, Bonell (1998) reviews the impacts of the conversion of forests to other land uses, under socio-economic pressure, in humid and semi-arid tropical regions – a subject that has been largely neglected so far. The paper describes controlled experimental studies in small catchments in the Amazon basin, Australia, India and the Sahel region of Africa. The effects of such changes in land use are compared with those of past natural changes in climate. The problem of the linkage between microscale hydrological results and the coarse resolution of atmospheric global climate models is recognized as a 'trans-science' problem. The paper also provides an extensive list of references.

In a broader context, some other projects are being conducted at an even higher level of abstraction:

- The IGBP project on Global Analysis, Interpretation and Modelling (GAIM). Its objectives are: '(1) To propose and facilitate experiments with existing models or to link sub-component models, especially those associated with IGBP Core Projects and with the efforts of the World Climate Programme. Such experiments will be focused on resolving interface issues and questions associated with developing and understanding the prognostic behaviour of key processes. (2) To clarify the key scientific issues faced in the development of Global Biochemical Models and the coupling of these models to General Circulation Models'.
- The Integrated Model for the Assessment of the Greenhouse Effect (IMAGE), developed by the Netherlands National Institute for Public Health and Environmental Protection (RIVM), consists of several submodels that can be used to study the consequences of various scenarios for emissions and land use. The RIVM has also launched the project Evaluation of Strategies to address Climate change by Adapting to and Preventing Emissions (ESCAPE).

1.5 SCOPE AND STRUCTURE OF THIS VOLUME

The mission of the Working Group on IHP-IV Project H-2.1 has been to compile information on the ongoing large-scale studies of the relationships between climate change (and variability) and hydrological regimes, their potential effects on water balance components, and to report on the results obtained so far.

This volume is relevant to the international community of scientists and practitioners of hydrology and climatology. It gives an overview of the climatic and hydrological changes that are already occurring around the world, and surveys the variety of methods that are being used to forecast future changes. Such a survey can of course never be complete, but it summarizes the areas for which information was available and accessible. It is clear that there are still large gaps that need to be bridged between the temporal and areal scales of resolution used in the modelling of atmospheric and hydrological phenomena and in the respective methods of study.

The preparation of this volume partly coincided with that of the second IPCC report, and some members of the Working Group and external assessors were involved in both. The two reports are quite different in nature, however. This report is restricted to the hydrological aspects of climate change and their consequences for water resources, and deals in particular with the various methods of 'trans-science' (hydrology and climate) research and their inherent problems. It therefore complements the IPCC report; there is little overlap and there are no major contradictions.

As the subject area is rapidly expanding, the prime contribution of the book is to provide a framework for synthesizing many diverse trains of thought, research and practice. This includes providing a connection between historical contemporary climate variability, future climate change impacts; and the various large-scale experiments for improving our understanding of climate regimes and hydrologic responses.

The effects of sea level rise on the hydrological cycle due to climate change may be considerable, as mentioned in Section 1.2.2, but are beyond the scope of this volume.

The structure of this volume is as follows:

Chapter 2 provides an overview of the work of the IPCC in the period 1988–94.

Chapters 3–7 deal with the conditions and situations in South America, North America, Europe, Africa, and Asia and Australia, respectively. Although the titles of these chapters are similar, their contents differ considerably, partly because the conditions and situations in each of the continents are different, and partly because the methods of study used are determined by the level of organization and the financial support allocated to such studies. These chapters were written by members of the Working Group, in some cases in collaboration with co-authors. All of the authors are either from, or have been involved in studies in the respective continents.

Chapter 8 is unique in that it gives a concise overview of the models that are being used in the evaluation of the impacts of climate change on hydrology.

Finally, Chapter 9 presents the conclusions and recommendations of the Working Group.

2 Climate change, hydrology and water resources: The work of the IPCC, 1988–94

I. A. SHIKLOMANOV

2.1 INTRODUCTION: PURPOSE AND OBJECTIVES OF THE IPCC

In the near future significant changes in the global climate can be expected, with an increase in mean air temperature of 3–4°C. For the northern temperate regions and high latitudes in particular, these changes are likely to affect a wide variety of physiographic features over vast regions, human living conditions, socio-economic structures and development, and natural ecosystems. The effects of a rise in global air temperature could be devastating: extensive melting of the polar icecaps and the resultant rise in sea level would impact coastal areas, and changes in atmospheric circulation would affect agricultural productivity and food supplies, as well as water resources in many countries.

In the early 1970s some scientists first warned of the possible impacts of human activities on the global climate (Budyko, 1972) and predicted a warming of 1.5–2°C in the coming decades. At that time, however, few climatologists supported the view that significant climate warming would result from increased concentrations of CO_2 in the atmosphere. On the contrary, it was widely believed that climate cooling would occur, and such a cooling trend was observed during the 1970s, according to data from the world meteorological network. Nevertheless, the first World Climate Conference (WMO, 1979) concluded that the human impacts on climate were serious, and that the problem required further study, but noted that science could not give a definite answer to the question of whether climate warming or cooling would occur.

The situation changed rapidly in the 1980s, when observation data showed a sudden rise in global air temperature. In 1985, the international conference on climate change and its impacts, in Villach, Austria, concluded that an increase in greenhouse gas concentrations during the first half of the next century would cause an unprecedented rise in mean global air temperature. The conference accepted the forecasts made on the basis of the recently developed general circulation models (GCMs), that a doubling of CO_2 (and minor greenhouse gases) in the atmosphere could cause global warming by as much as 1.5–4.5°C by the year 2030. The conference therefore called for further studies to assess the likely negative impacts of global warming on social, economic and ecological systems, and to formulate measures to prevent and/or to mitigate them.

By the late 1980s the problem of climate change and its possible impacts had become an issue of global concern. Noting the anxiety of scientists, international organizations and some national governments, the World Meteorological Organization (WMO) and the UN Environment Programme (UNEP) established in 1988 the Intergovernmental Panel on Climate Change (IPCC), headed by Professor Bert Bolin. The Panel was charged with the following: (a) to assess the scientific information on various aspects of climate change, such as emissions of major greenhouse gases and the resultant modification of the Earth's radiation balance; (b) to use that information to evaluate the environmental and socio-economic consequences; and (c) to formulate realistic response strategies.

The IPCC began its task by establishing three working groups, with the following responsibilities: WG-I was to compile all the available scientific information on climate change; WG-II was to assess the potential environmental and socio-economic impacts of climate change; and WG-III was to formulate response strategies. A special committee was also set up to promote, as quickly as possible, the full participation of developing countries (IPCC, 1990a).

2.2 ORGANIZATION OF THE WORK OF THE IPCC

The work of the IPCC has proceeded in several stages. The first stage began in 1988, and was completed in October 1990 with the submission of reports to the Second World Climate Conference in Geneva (IPCC, 1990b). In the second stage,

1991–92, a supplementary report was compiled containing updated data (IPCC, 1992). In 1993 the work of the IPCC was extended to include more specialists from different countries. The preliminary findings of the third stage of the IPCC's activities were considered in late 1994, and the final report was published in summer 1995 (IPCC, 1995).

At each stage, the IPCC's WG-II has been concerned with assessments of the potential impacts of climate change on hydrological regimes and water resources, including problems of water management, consumption and utilization, water availability, and the protection of water supplies. At the first stage, WG-II was headed by co-chairmen Dr G. Lins (US Geological Survey), Professor I. Shiklomanov (Russian State Hydrological Institute), and Dr E. Stakhiv (US Corps of Military Engineers). Dr M. Kara (Algeria) was nominated to represent the developing countries. These specialists prepared a chapter on hydrology and water resources for the report of WG-II (Shiklomanov et al., 1990), submitted a report to the Second World Climate Conference (Lins et al., 1991), and later published a supplementary report on the impacts of climate change on hydrology and water resources (Stakhiv et al., 1992).

During the second stage, two groups of specialists were set up to assess the impacts of climate change on freshwater systems, and to prepare two chapters for the IPCC's final report. The first group, headed by Professor H. Lang (Switzerland), with members B. Bates (Australia), Chunzhen Liu (China), S.K. Mugera (Kenya) and O. Starosolszky (Hungary), was to prepare Chapter 10, on hydrology and freshwater ecology. The second group, headed by Professor Z. Kaczmarek (Poland), with members N. Arnell (UK), E. Stakhiv (USA), K. Hanaki (Japan), G. Mailu (Kenya) and L. Somlyódy (Hungary), prepared Chapter 14, on water resources management. These materials were used in the preparation of the present report.

2.3 MATERIALS

The task of the IPCC was not to conduct research on the problem in different regions, nor to prepare long-term scientific programmes and outlines, but to gather together the results of research conducted in various countries. The Panel was then to generalize the results from the scientific viewpoint, to prepare proposals for future research, to develop practical recommendations, and to bring the results to the attention of international organizations, decision makers and the public. To meet its objectives, the IPCC needed to consider as much material as possible from different physiographic regions, reflecting the different socio-economic conditions in all continents, in both developed and developing countries.

The materials used in the preparation of the IPCC's reports on hydrology and water resources were as follows:

- Special surveys summarizing the research of individual authors in their own countries, including up-to-date unpublished materials. Detailed reports were also prepared for some other countries, such as Belgium, Germany and New Zealand.

- Materials prepared by international organizations, and the proceedings of international meetings, such as the symposium on the impacts of climate variability on hydrological regimes and water resources (Vancouver, Canada, 1987), the US–Canada symposium on the effects of climate change on the Great Lakes basin (1988); the conference on climate and water (Helsinki, Finland, 1989); and the conference on the effects of climate change on the environment and society (Japan, 1991). Extensive information was also obtained from the proceedings of national symposia in Australia, the USA, etc.

- Reports published in the previous ten years by various national and international organizations; these were the most accessible sources of information for the IPCC reports, which cite more than 600 references.

The IPCC also used the reports of studies conducted for river basins and regions in the following countries (in alphabetical order): Algeria, Argentina, Australia, Bangladesh, Belgium, Brazil, Canada, Chile, China, Denmark, Finland, France, Germany, Hungary, India, Indonesia, Israel, Japan, Kenya, Nepal, The Netherlands, New Zealand, Norway, Peru, Poland, Romania, Russia (former Soviet Union), Senegal, Sweden, Switzerland, Thailand, UK, Uruguay, USA and Venezuela. The continents of Australia, Europe and North America have been most completely investigated, whereas Africa and some regions of Asia and South America have been only poorly studied.

2.4 SCENARIOS AND METHODOLOGICAL APPROACHES

Until recently, forecasts of anthropogenic climate change have been unreliable, so that scenarios of future climate conditions have been developed to provide quantitative assessments of the hydrological consequences in some regions and/or river basins. These scenarios can be classified into three groups: hypothetical scenarios, climate scenarios based on

GCMs, and scenarios based on reconstructions of warm periods in the past (paleoclimate reconstructions).

Hypothetical scenarios are based on arbitrary specified changes in future climate characteristics. In most of these scenarios it is accepted that air temperature will rise by 0.5–4°C and that precipitation will change (rise or fall) by 10–25%. Some authors also specify hypothetical changes in evaporation. In some reports scenarios have been specified proceeding from analyses of changes in climate characteristics that occurred in the past during particularly warm or cold periods of hydrometeorological observations. Using hypothetical scenarios that are not related to a particular time in the future, it is possible to assess the probable responses of a river basin to certain changes in climate parameters. These hypothetical scenarios were applied mainly in the early stage of research in the 1980s, but are now only rarely used.

Climate scenarios are based on the results of the computations of GCMs, several of which have been developed. The GCMs most often used to assess the hydrological consequences of climate change are those of the US Geophysical Fluid Dynamics Laboratory (GFDL), the Canadian Climate Centre (CCC), and the UK Meteorological Office (UKMO). Other widely applied models include the GISS, OSU and NCAR models (USA), and the model developed at the Max Planck Institute (Germany). The first three models were recommended by the IPCC's WG-I (IPCC, 1990a); they provide the highest resolution, and consider the case of a doubling of atmospheric CO_2 concentrations by the year 2050. Recently, some GCM scenarios have been generated based on CO_2 increases of 10%, 20% and 30%.

GCM-based scenarios can provide details of changes in regional climates (monthly data on air temperature, precipitation, and other parameters) for almost all areas on Earth. At present, however, the results of such scenarios are not very reliable; different GCMs often provide different and even contradictory results even for the same region, on precipitation in particular. This is very important for assessments of hydrological consequences. The results of GCM-based and hypothetical scenarios for the same regions are also often different (at least in the case of precipitation).

Paleoclimate scenarios are based on the use of climates in the past, when concentrations of CO_2 in the atmosphere were higher than at present, as prototypes of future climate conditions. For example, Russian climatologists (Hydrometeoizdat, 1987; Budyko, 1988) showed that the so-called Holocene optimum (about 5000–6000 BC) might be regarded as a prototype of global warming by 1°C (in about 2000–2005); the last interglacial period 12 500 years ago (the Mikuliskian interglacial) as a prototype of global warming by 2°C (2020–2025); and the so-called climate optimum of the Pliocene of several million years ago, when mean air temperatures were 3–4°C higher than today, might serve as a prototype of climate conditions in the more distant future (2040–2050).

Charts of the approximate distribution of the expected changes in air temperature and precipitation are available for the northern and southern hemispheres for each of these prototypes. These charts were obtained using various indirect methods and are accepted as the basis for assessments of the hydrological consequences of climate change. The most detailed charts were prepared for the former Soviet Union, and have been widely applied by Russian hydrologists. Such assessments have been made for river basins in Russia, as well as in South America (Budyko *et al.*, 1994), and as a first approximation for the continents on average (Shiklomanov and Babkin, 1992). Scenarios based on paleoclimate data have also been developed for New Zealand (Griffiths, 1989).

The use of paleoclimate prototypes is very promising in that they allow consideration of possible variations in water resources within definite time intervals in the future, including the next 10–15 years. At the same time, this method presents some evident difficulties and constraints associated with, first, conventions regarding accepted climate prototypes in the remote past and future, and second, by the lack of reliable paleoclimate data for many regions. An assessment based on a paleoclimate reconstruction may therefore be regarded as just one possible version of the climate situation in the future.

Where climate scenarios are available, the quantitative assessments of the hydrological consequences of climate change analysed in the IPCC reports use the following sets of methods:

1. statistical dependencies between runoff and meteorological characteristics for the periods of observation;
2. long-term water balance for the periods of observation;
3. direct use of GCM computations; and
4. deterministic hydrological models for river basins for short time intervals.

The first and the second sets of methods have been widely applied by WG-II, especially during the first stage of research in the 1980s, because of their simplicity and the low input data requirements. However, the reliability of analyses using these approaches is low, especially with respect to changes in hydrological regimes over short time intervals (months, 10-day periods). It is difficult to apply these estimates to evaluate extreme characteristics of river runoff, so that statistical dependencies are rarely used. The long-term water balance method can provide only approximate assessments of changes in mean annual runoff within large river basins and regions. The accuracy of these assessments depends on the reliability of forecasts of changes in mean annual precipitation and on the methodology used to assess evapotranspiration.

In the application of GCMs for direct hydrological purposes, the model results are far from consistent, and for some important hydrometeorological characteristics and regions they are often contradictory. This is because in most GCMs the level of resolution is low (for hydrology), and they describe hydrological processes schematically, simplifying them greatly. Nevertheless, the approach is promising and investigations in this field should continue.

The use of deterministic hydrological models is preferable for current conditions (for an overview of such models, see Chapter 8). With this approach it is possible to investigate explicitly the cause–effect relations in the climate–water resources system, to evaluate the vulnerability of river basins to changes in climate parameters, to compute changes in runoff regimes under different physiographic conditions using available regional climate forecasts, and to identify the implications for future water projects. Various hydrological models have been used in these assessments, including the well-known basin models used in hydrology, and models developed for studying the effects of changes in climate parameters on hydrological characteristics.

In accordance with the IPCC's conclusions, it is reasonable to adopt the following scheme for assessing the impacts of climate change on hydrological regimes and water resources. The results of regional climate assessments obtained from GCMs or paleoclimate prototypes serve as the inputs to hydrological models of river basins; in turn, the output of the hydrological models, together with the results of climate assessments, serve as inputs for models of water consumption, water use, and water resources management.

2.5 POSSIBLE CHANGES IN HYDROLOGICAL REGIMES AND WATER QUALITY

In the published results of analyses of the potential impacts of climate change on the hydrological characteristics of water bodies, the levels of accuracy vary considerably and the results are rarely compatible. This is due to the variety of input data used in climate scenarios and the different methodologies applied. Nevertheless, on the basis of materials submitted to the IPCC it is possible to identify some general trends for a number of world regions.

2.5.1 Temperate regions

In the temperate regions of northern and eastern Europe, northern Asia and North America the largest proportion of annual runoff is derived from spring snowmelt. Most of the studies of changes in annual and seasonal runoff, in streamflow distribution throughout the year, and in extreme runoff parameters due to global warming, have been made for these regions. Climate scenarios based on GCMs and on paleoclimate prototypes have been used to assess hydrological consequences.

For the former Soviet Union, for example, Shiklomanov and Lins (1991) showed that with a temperature increase of $1°C$, total annual river runoff could increase by about 7%, or 280–320 km^3/year, and with an increase of $2°C$ the runoff of large river systems could increase by 10–20%, or 700–800 km^3/year. For river basins dependent on spring snowmelt, however, climate change is likely to have a greater effect on the streamflow distribution throughout the year than on mean annual values. In a study of the effects of global warming on annual and seasonal runoff of the largest river systems (Volga, Dnieper and Don), using available data, Shiklomanov and Lins (1991) found significant changes in seasonal (compared with annual) runoff, i.e. sudden increases in winter runoff and reductions in spring snowmelt runoff, due to more intense snowmelt in winter.

Similar conclusions have been drawn for regions with comparable physiographic conditions (Belgium, Canada, Poland, Scandinavia, Scotland, etc.). For example, the GISS GCM, under a $2 \times CO_2$ scenario, has been used to assess changes in runoff for 12 basins in Finland draining areas of 600–33 500 km^2. The results show increases in annual runoff of 20–50% and a more even streamflow distribution throughout the year in some basins. The maximum spring snowmelt discharges tend to decrease by up to 55%, while minimum monthly discharges tend to increase by 100–300% (Stakhiv et al., 1992). Similar changes in seasonal runoff (increased winter runoff, reduced spring snowmelt runoff), although less pronounced, have been obtained for river basins in Belgium, Norway, Poland and Switzerland.

For a large part of Canada, global warming could lead to an increase in mean annual river runoff and a more even streamflow distribution throughout the year, except in the Great Lakes basin, where a decrease in river runoff is likely. These changes may alter the water balance and the lake level regime, and could also threaten the economy and the ecology of the region.

According to Gleick (1987a, 1988), Lettenmaier and Gan (1990) and Lettenmaier and Sheer (1991), significant changes in seasonal and monthly runoff due to global warming can be expected in the river basins of California, where runoff is derived from snowmelt in the mountains.

For New Zealand, changes in annual runoff due to global warming have been assessed on the basis of paleoclimate scenarios. In this region about 8000–10 000 years ago the air temperature was 1.5°C higher, and annual runoff was much higher than it is today.

In summary, the materials submitted to the IPCC for temperate regions indicate that global warming will tend to increase mean annual runoff. In regions where runoff is derived mainly from spring snowmelt, significant changes in seasonal and monthly runoff can be expected. In this case the streamflow distribution throughout the year would be more vulnerable to changes in air temperature than in annual precipitation.

2.5.2 Arid and semi-arid regions

The IPCC investigations show that global warming could affect the hydrological regimes of arid and semi-arid regions even if the changes in climate characteristics are minor. For example, an increase in mean annual air temperature by 1–2°C and a 10% decrease in precipitation could reduce annual river runoff by up to 40–70%. These conclusions were based on estimates for river basins in areas of water deficit in the USA, Canada, Australia, Russia, Africa and South America. Table 2.1 shows the results of some assessments of climate change impacts on annual river runoff for areas of water deficit. Similar estimates have been obtained for other river basins in the USA, Russia, Australia and northern China. Meanwhile, according to Stakhiv et al. (1992), the earlier conclusions that river basins in arid and semi-arid regions would be vulnerable even to slight changes in climate characteristics may be overstated.

In analyses of the hydrological consequences of global warming in arid regions, particular attention has been paid to the Sahel, where the severe droughts of recent decades have already caused critical situations in many countries. In a multidisciplinary analysis of the effects of climate change on the water resources of the Sahel submitted to the IPCC, Sircoulon (1990) noted the vulnerability of river basins to changes in climate parameters, and the inadequacy of most water management systems. There is a large degree of uncertainty about the water resources of the Sahel, how-

ever, since the analyses of the future climate based on paleoclimate reconstructions and on GCMs give quite contradictory results. According to the paleoclimate scenarios, a slight increase in air temperature could increase precipitation in the Sahel, resulting in an increase in water resources. In contrast, according to the three GCMs (IPCC, 1990a), assuming a doubling of CO_2, an increase in air temperature of 1–2°C would be accompanied by a reduction in winter precipitation of 5–10% and an increase in summer precipitation of up to 5%. Under this scenario an increase in river runoff is rather unlikely.

In regions of water deficit, with different values of the expected changes in river runoff, it is evident that annual runoff is much more sensitive to changes in precipitation than in air temperature.

Most assessments of changes in annual runoff due to global warming (including those cited in Table 2.1) do not take account of the possible direct effects of higher CO_2 concentrations on plant physiology and transpiration capacity (at higher CO_2 concentrations transpiration tends to decrease). However, such factors were considered by Idso and Brazel (1984) in a study of five river basins in Arizona (USA), and by Aston (1984) for watersheds in Australia. Their results differed widely from those of other scientists. With a doubling of CO_2, the annual runoff of US rivers increased by 40–60%, and of Australian rivers by 60–80%. Many other scientists have also noted the potential direct influence of CO_2 on transpiration and evapotranspiration, and consequently also on runoff. The first IPCC report noted that this problem required further investigations, since it is likely to affect the reliability of quantitative assessments of the hydrological consequences of global warming. The report also noted that although this factor is important in detailed investigations, it is unlikely that it will be as significant as stated by the above authors because evapotranspiration from land is determined mainly by energy factors.

This conclusion was confirmed in later studies. For example, Easterling et al. (1991) found that the direct effect of higher CO_2 concentrations would be to increase precipitation by 35–450 mm, leading to a 10% reduction in irrigation water requirements. Experiments in Japan (CGER, 1993) showed an increase in productive biomass and a decrease in water consumption with increasing CO_2 concentrations, but this depends to a large extent on the type of vegetation. Kimball et al. (1993) found that an increase in CO_2 of up to 550 parts per million would reduce the evapotranspiration of winter wheat by about 11%. It can be assumed that the irrigation water requirements for wheat and probably for

Table 2.1. *Results of assessments of impacts of climate change on annual river runoff (basins and areas of water deficit).*

Region and basin	Climate change scenario		Change in annual runoff (%)	Reference
	Temperature (°C)	Precipitation (%)		
Mean for seven large basins in the western USA	+ 2	− 10	− 40 to − 76	Stockton and Boggess (1979)
Hypothetical arid basins	+ 1	− 10		Němec and Schaake (1982)
Colorado River, USA	+ 1	− 10	− 50	Revelle and Waggoner (1983)
Peace River, USA	+ 1	− 10	− 50	Klemeš and Němec (1983)
River basins in Utah and Nevada, USA	+ 2	− 10	− 60	Flaschka et al. (1987)
River basins in the steppe zones of European Russia	+ 1	− 10	− 60	Shiklomanov (1989)

some other crops would be slightly reduced with higher CO_2 concentrations (IPCC, 1995).

2.5.3 Humid tropical regions

In the IPCC reports the impacts of global warming on water resources in humid tropical regions have been assessed on the basis of studies of two river basins in Venezuela, for the Uruguay River in Uruguay, for the Lower Mekong basin in Vietnam, for river basins in Sri Lanka and Indonesia (Stakhiv et al., 1992), as well as for South Asia and central India (IPCC, 1995). GCM climate scenarios (GISS, UKMO, GFDL, OSU) were used in each case, except for central India and Venezuela, where river basin responses were estimated from hypothetical scenarios. The quantitative conclusions differ considerably.

For example, in a study of the Uruguay River, Tucci and Damiani (1991) compared the results of the GISS, GFDL and UKMO GCMs for the case of $2 \times CO_2$. The authors concluded that all three models underestimated the amounts of precipitation, but the GISS model better reflected the historical situation. The results from the hydrological model with the GCM scenarios were as follows: runoff decrease by 11.7% (GISS), runoff increase by 21.5% (GFDL), and runoff decrease by 6.4% (UKMO). The differences were even greater for other river basins; moreover, the results for changes in runoff depend mainly on the assumed precipitation. Many authors have noted that GCMs not only provide quite different forecasts of future precipitation, but they are also often unable to reproduce the present climate conditions, for monsoon regions in particular.

2.5.4 River basins in mountainous regions

Recent assessments for mountain river basins in the Alps and for a small watershed in Nepal (IPCC, 1995) show reductions in the duration of snow cover, some increases in annual runoff, and significant changes in streamflow distribution throughout the year, due mainly to higher air temperatures. For example, Bultot et al. (1994) used a hydrological model for a small watershed in the Swiss Alps, and found that with a 1% increase in air temperature, snowmelt flood runoff would be reduced, while floods due to rain would occur more often, because the proportion of rain in the total precipitation would be greater. A study of the French Alps showed that the responses of small watersheds in mountainous areas to climate change would depend greatly on the elevation (IPCC, 1995).

In its most recent report the IPCC (1995) describes specific features of changes in hydrological regimes in mountainous areas due to global warming. The report also notes that there are large uncertainties, due to factors such as orographic effects, specific features of snow and ice storage and melting, the effects of altitude on climate parameters, and the difficulties involved in determining precipitation and evaporation in mountainous areas.

2.5.5 Large river systems

Hydrologists from many countries have investigated the impacts of climate change on large river systems in Europe (the Danube, Dnieper, Don, Rhine, Volga and Warta river systems); in North America (the Colorado,

Sacramento, Mississippi–Missouri, Delaware and Rio Grande); in South America (Amazon, La Plata, Paraná, Paraguay and Uruguay); in Africa (the Congo, Zambezi, Senegal, Oranje and Nile); in Asia (the Mekong, Indus, Ganges, Yellow and Yangtze); and in Australia (the Murray–Darling River system).

In such a wide range of studies, it is to be expected that the degree of detail varies, and that the results will depend on the chosen future climate scenario and the physiographic features of the region considered. For example, Miller and Russell (1992) discussed the impacts of global warming on the runoff of 33 of the world's largest rivers, using the GISS GCM, under the $2 \times CO_2$ scenario. They found that in all rivers at high northern latitudes an average increase in runoff of 25% can be expected, due to increased precipitation. Simultaneously, lower runoff maxima can be expected in river basins at low latitudes (−43% for the Indus basin, −31% for the Niger, and −11% for the Nile) due to the combined effects of increased evapotranspiration and reduced precipitation. The authors noted some large discrepancies between the observed and computed (by GCMs) climate parameters for the present climate. The results for river systems at high northern latitudes are in a good agreement with those obtained in Russia, Hungary and the Netherlands using GCM-based and paleoclimate scenarios. The latter have shown, for example, large increases in annual or winter runoff for the Danube, Dnieper, Don, Rhine, Volga, Yangtze, and other large rivers of eastern Europe and northern Asia with global warming. For large river basins at low latitudes in areas of water deficit, the paleoclimate scenarios provide contrasting results – significant increases in runoff with global warming – because, unlike GCMs, they assume greater increases in precipitation in arid and semi-arid regions with only a slight increase in mean air temperature.

2.5.6 Lake water balance and water levels

Climate variability is likely to affect the hydrological regimes of many large lakes and their basins, leading to changes in all components of the water balance of lakes (precipitation, evaporation, inflows and outflows), as well as lake levels and thermal regimes, and ecosystems. The nature of these changes would be different in endorheic and exorheic lakes.

The IPCC reports provide only limited information on the impacts of climate change on the hydrological regimes of lakes. The Great Lakes of North America, Lake Kariba in Africa, and for the Caspian Sea (Shiklomanov and Lins,

1991) have been studied in greatest detail. Data are also available on lakes whose water levels are likely to fluctuate over the long term regardless of climate change: in the Antarctic, in the Mackenzie delta in Australia, as well as Lake Kinneret (Middle East), Lake Michigan (North America) and Lake Titicaca (South America) (IPCC, 1995).

The Great Lakes contain more than 8000 km^3 of water, cover an area of 246 000 km^2, and drain an area of 766 000 km^2. All of the lakes are connected with each other and with the Atlantic Ocean. Hydrometeorological data have been recorded for more than 100 years, during which time the maximum range of water level fluctuations has been little more than 2 m. Assessments of variations in climate parameters due to global warming in the Great Lakes watershed area have been made using various GCMs. According to one of the first studies (Sanderson and Wong, 1987), assuming a doubling of CO_2, with a rise in air temperature of 4.4°C and an increase in precipitation of 6.5%, hydrological models give reduced river runoff in the Great Lakes basin and a redistribution of streamflow throughout the year: the maximum runoff occurs much earlier (by up to a month) compared with natural, undisturbed conditions. Water levels in all of the lakes can be expected to fall by 20–60 cm, the range of monthly level fluctuations would be a little wider, and the outflows from Lakes Michigan, Huron and Erie would be reduced by 12–13%. Such fluctuations in lake levels and outflows are unlikely to occur, however, because new lake control measures are likely to be introduced in response to any new situation. Later assessments and the outputs of other GCMs provide quite different results, some of which demonstrate significant increases in runoff in river basins in the future (Waterstone et al., 1995).

Global warming could have a dramatic effect on the endorheic Caspian Sea. A 2–3°C increase in air temperature would lead to increased precipitation, and hence to greater discharges from the rivers such as the Volga that drain into the sea (Shiklomanov and Lins, 1991). Results such as these could be of great relevance in the formulation of policies and measures to deal with the already acute problem of the fluctuating level of the Caspian Sea.

2.5.7 Extreme runoff values and other dynamic characteristics

One of the most important objectives of studies of the hydrological impacts of global warming, is to estimate possible changes in extreme river runoff characteristics: both the maximum discharges and flood volumes, and minimum flows for limited periods.

The forecasting of extreme river runoff parameters in response to climate change is one of the most challenging problems of engineering hydrology. Until recently, most water projects have been designed with the assumption that hydrometeorological conditions will remain the same in the future; i.e. that the observation data collected over the years will remain valid throughout the period of operation of the project. All of the applied statistical methods that are used to assess the size of a planned structure, its capacity and its resilience to extreme events, are based on this assumption. However, climate change will undoubtedly affect extreme runoff parameters, so that the use of the simplest statistical methods will be unjustified from a scientific viewpoint. The general approach to the design of water projects needs to change fundamentally. Even greater problems are likely to arise for hydraulic structures already in operation.

The problem of obtaining reliable quantitative estimates of maximum discharges with global warming is, however, one that can not be solved at present, or at least is much more complicated than assessing changes in annual and seasonal runoff. Maximum discharges depend on the intensity and duration of precipitation during individual storms or periods of rapid snowmelt, and minimum water discharges occur after extended periods without rainfall. Since these meteorological events can not be predicted using GCMs or paleoclimate prototypes in particular, no reliable estimates can be made of possible changes in extreme runoff parameters. Here, Klemeš (1992) is probably correct in his suggestion that in the computation of extreme hydrological parameters under conditions of global warming we are witnessing even greater uncertainty.

Nevertheless, climatologists such as Houghton *et al.* (IPCC, 1992) assume that under warmer conditions the hydrological cycle will become more intense, stimulating rainfall of greater intensity and longer duration, causing longer periods of floods and droughts. Several hydrologists have reached similar conclusions in their assessments of changes in extreme characteristics under global warming, assuming that they are proportional to changes in annual, seasonal or monthly runoff. For example, increases in maximum floods are likely to occur in northwestern USA and northeastern Canada, as well as higher winter maximum and lower summer minimum discharges in California. Significant increases in runoff maxima may be expected in northwestern Europe, and increases in winter runoff minima and reduced spring snowmelt in eastern Europe and western Siberia. According to Griffiths (1989), the

frequency and intensity of floods in New Zealand will increase if temperatures rise by 1.5°C. Yamada (1989) analysed long-term variations in air temperature and precipitation in Japan, and found that global warming could intensify rainfall on the one hand, and reduce the amount of precipitation during low-water periods on the other. Thus, in this particular case, global warming could cause an increase in maximum rainfall–runoff and a decrease in minimum discharges. Liu *et al.* (1995) found that global warming would intensify the variability of the Asian monsoon in northern China, increasing the frequency of floods and droughts. The frequency of floods in humid areas of southern China would also increase. In Australia (Australian Bureau of Meteorology, 1991) there would be greater variability in the extreme parameters of precipitation and river runoff. Model results show that the maximum precipitation would tend to increase in summer, and minimum precipitation would decrease in winter.

In summary, global warming is likely to result in greater changes in extreme runoff parameters than in mean annual and seasonal parameters, particularly in small and medium-sized watersheds. Most researchers expect that the frequency of floods will increase considerably (IPCC, 1995), although the amount of such increases is uncertain and would differ from region to region.

The hydrological consequences of global warming include changes in river runoff and the water balance, as well as in other parameters such as total water availability, water levels, and maximum and minimum discharges, which in turn would affect rates of erosion on slopes and in river channels, increasing sediment concentrations in water flows, sediment yields and in channel processes. In China, for example, the regional effects of climate change on sedimentation in the Yellow River basin have been studied using GCM scenarios and conceptual hydrological models (Bao, 1994). The results indicate that runoff would tend to decrease, while sedimentation would increase due to the increased storm intensity during the flood season.

Sea level rise and changes in river runoff will cause the flooding of many low-lying coastal areas, and intensify the erosion of riverbanks and seashores. Changes in the processes of delta formation would then contribute to the salinization of the lower reaches of rivers and estuaries due to saltwater intrusion. These problems have not been adequately addressed in the reports submitted to the IPCC. Such assessments require multidisciplinary research that takes into full account detailed forecasts of changes in regional climate characteristics and water regimes under various physiographic conditions.

2.5.8 Water quality, chemical and biological processes

The deterioration of water quality is likely to be one of the most serious hydrological consequences of global warming. This problem was not discussed during the first stage of the IPCC's activities due to insufficient information, although it is considered in several sections of the most recent report (IPCC, 1995, Chapter 14 on water resources management).

Assessments of future water quality are complicated because water quality characteristics depend on a number of human factors (such as the existence of point and scattered sources of contamination, land use practices), and on such 'natural' factors as air temperature, precipitation, solar radiation, wind, CO_2 concentrations, etc. Climate change will affect all of these factors, which in turn will affect water quality directly, and in many cases indirectly (through hydrological and hydraulic parameters).

Climate change is likely to affect polluted bodies of water more significantly than clean ones. Thus the deterioration of water quality may be more serious in developing countries where pollution control regulations are less stringent than in the industrialized countries. Higher CO_2 concentrations in freshwater would alter the system of inorganic carbon equilibrium, leading to lower pH. In a study of Lake Balaton in Hungary, Szilugyi and Somlyódy (1991) found that a higher carbon separation in water would probably contribute to higher water oxidizability compared with the effect of acid rain. A doubling of CO_2 would lead to a reduction in pH by 0.2, thus increasing the salt content and the hardness of lake water. The increase in inorganic carbon may stimulate primary production, leading to eutrophication.

Changes in hydrological and hydraulic parameters could have considerable effects on water quality. Reduced river runoff would mean less dilution, thus increasing salinity and pollution concentrations; expensive projects would be required to reduce the effects of such processes. The smaller volumes of water could affect a number of biological processes, including the balance of dissolved oxygen, nitrification and denitrification, as well as eutrophication, and the higher temperatures could affect the carbon, nitrogen and phosphorus cycles in water. The effects of these interrelated processes can be analysed by multipurpose models of the nutrition cycle if it is assumed that the temperature would reflect climate variations.

Changes in the ice cover on water bodies due to global warming could also affect water quality. Studies of the Great Lakes of North America, and of the lakes in Finland, have shown that reductions in the duration of ice cover by two or three times could occur with a doubling of CO_2. Other important impacts of global warming include changes in the thermal regimes and the stratification of freshwater lakes. Changes in water temperature, ice cover and stratification may induce significant changes in freshwater ecosystems and water quality.

In summary, clean freshwater lakes are likely to be little affected by climate change, but if the water is polluted, additional changes could occur that can be estimated only with a high level of uncertainty. These effects could be analysed using available models of hydrothermal regimes and water quality (IPCC, 1995).

The latest report of the IPCC WG-II (1995) details the potential effects of variations in temperature and hydrological parameters on river and lake ecosystems, plankton, fish stocks, accumulation of pollutants, ice regimes and snowmelt. These problems are now being addressed in many areas of hydrological research, in order to determine the effects of global warming on freshwater ecosystems. However, these issues are not directly relevant to quantitative assessments of the likely impacts of climate change on hydrology. Few such assessments consider freshwater ecosystems under different physiographic conditions, and all of the available results are rather uncertain, for a number of reasons. First, freshwater ecosystems are affected directly by human activities (rather than climate), and it is difficult to evaluate what these impacts will be in the future. Second, ecosystem changes are secondary to changes in water regime parameters (water levels, discharges and volumes), and the assessment of these parameters is also rather uncertain. Third, changes in freshwater ecosystems, unlike water regime parameters, may become evident only over the course of many years. All of these factors require time-consuming, multidisciplinary research, as well as detailed and reliable long-range forecasts of regional climate change.

2.6 CLIMATE CHANGE, WATER AVAILABILITY AND WATER RESOURCES MANAGEMENT

In assessing the hydrological impacts of global warming, the issue of water resources management has been an important component of the work of the IPCC's WG-II at every stage. The first IPCC report (Shiklomanov et al., 1990) presented a survey of case studies of the possible effects of climate change in terms of water management, particularly in vulnerable regions such as the basins of large water bodies (Great Lakes); the Sahel; regions of intensive agriculture (the South Platte River, USA, and the Murray–Darling basin, Australia); expanding urban areas (the Delaware River basin, USA); and regions where water supplies are derived from snowmelt (the Sacramento–San Joaquin Rivers, USA). For these regions, changes in hydrological regimes under various climate scenarios have been determined, but only approximate assessments have been made of the possible impacts on the availability of water for domestic use, agriculture, power generation, navigation, industry, recreation, natural ecosystems, etc.

In the supplementary report to the IPCC's impact assessment, particular attention was paid to the effects of global warming on water use and water resources control (Stakhiv et al., 1992). The report summarized the conclusions of two conferences held in 1991 and 1992, which reflected the experience, physical features and socio-economic situation in the USA. Together with possible economic and ecological consequences of water management strategies, the report considered adaptive measures that might be taken to mitigate the impacts of climate change.

The most recent IPCC report (IPCC, 1995) contains a wide range of new research results, although it is noted that there was insufficient new information to warrant a substantial revision of the previous conclusions. However, various aspects of water management in relation to physiographic and socio-economic conditions are considered in greater detail. The report also discusses the effects of climate change on water supplies in various river basins, and provides estimates of changes in water requirements for agriculture, industry, power generation, navigation, recreation, and domestic and other uses. The report discusses the conclusions of recent investigations and the issue of water quality, as well as the management implications of various adaptation options.

2.6.1 Impacts of climate change on water availability

The IPCC report classifies the numerous studies of the effects of climate change on water availability into three groups. The first group of studies addresses potential changes in water availability directly by model estimates of annual and monthly water balances; the second group deals with the responses of hypothetical water supply systems (i.e. individual reservoirs); and the third group deals with possible changes in actual water systems. Some studies consider particular reservoirs or underground water systems, whereas others consider multipurpose water supply systems with existing standards of control. Most of the studies were conducted assuming conditions in the USA; some refer to the UK, Greece and the former Soviet Union (Dnieper basin),

but there have been few studies of the river basins of Africa, Asia and South America.

These studies indicate that a doubling of CO_2 could have significant impacts on water supply systems. These impacts may be positive or negative, depending on the climate scenario adopted, the water management sector concerned, and physiographic conditions. Many of the negative economic and environmental impacts could be minimized by the introduction of adaptive measures.

The IPCC report considers the changes in water availability at global and regional scales. It analyses the specific (per capita) water availability by the year 2050 for 16 countries under conditions of a stationary climate and under three GCM scenarios. These data are then compared with current (1990) levels of water use. Changes in water resources due to global warming for each country were estimated using the method of Kaczmarek (1990b) with air temperature and precipitation data. Table 2.2 shows the effects of two factors – population growth and climate change – on water availability according to the three GCM scenarios.

From this analysis the authors conclude that for all developing countries with high population growth rates, the per capita water availability tends to decrease irrespective of the climate scenario. However, for many countries the various model results are inconsistent; this highlights the difficulty in formulating appropriate adaptive water management strategies if they are to be based on current predictions of the effects of climate change. It may be expected that factors other than climate will stimulate projects to ensure the availability of reliable water supplies in many parts of the world, and that climate change may introduce still greater uncertainty in the development of methods for water resources control.

Still greater problems arise in assessing the availability of water in the countries and regions that share international river basins (Danube, Mekong, Nile, etc.). Changes in runoff may lead some countries to break treaties on the shared use of water, which could result in international disputes.

2.6.2 Climate change and water demand

Global warming could result in changes in water demand, in the redistribution of water resources throughout the year in many regions, in the structure and in the nature of water consumption, and more intense disputes among different water users.

The results of assessments of possible changes in irrigation water requirements for various world regions are ambiguous and depend on the scenarios applied, on the methods of

Table 2.2. *Water availability (m³/year per capita) in 2050 according to the three climate scenarios (data from Kaczmarek, 1990b).*

Country	Present use (1990)	Present climate (2050)	GFDL (2050)	UKMO (2050)	MPI (2050)
Cyprus	1280	770	470	180	1100
El Salvador	3760	1570	210	1710	1250
Haiti	1700	650	840	280	820
Japan	4430	4260	4720	4800	4480
Kenya	640	170	210	250	210
Madagascar	3330	710	610	480	730
Mexico	4270	2100	1740	1980	2010
Peru	1860	880	830	690	1020
Poland	1470	1200	1160	1150	1140
Saudi Arabia	310	80	60	30	140
South Africa	1320	540	500	150	330
Spain	2850	2680	970	1370	1660
Sri Lanka	2500	1520	1440	1600	4900
Togo	3400	900	880	550	700
UK	2090	1920	1810	1510	1750
Vietnam	5640	2630	2680	2310	2760

computation, and on physiographic conditions. In general, however, the expected changes are not great. For example, McCabe and Wolock (1992) examined the situation in some US regions on the basis of the Thornthwaite method, and found some increases in irrigation water demand even with a 20% increase in precipitation. The authors concluded that irrigation water requirements are more dependent on temperature than on precipitation.

However, contrasting results have been obtained at the State Hydrological Institute for the southern regions of the former Soviet Union. Georgievsky *et al.* (1993) used a specially developed energy–water balance method to determine evaporation levels, and found that changes in precipitation had a greater effect on demand than changes in air temperature. Within the former Soviet Union, technological improvements can be expected to reduce irrigation water losses considerably; in many regions even minor improvements in technologies could have a greater impact on water demand than changes in climate parameters. This is particularly the case with water supplies for industrial, domestic and other users, whose levels of demand are often dependent on the technologies used rather than on climate conditions.

The IPCC report notes that global warming would seriously affect fisheries, particularly freshwater fish reproduction. Some fish species are highly sensitive even to minor changes in water temperature.

2.6.3 Lessons from case studies: Management implications and adaptation

From numerous studies of the possible consequences of climate change for water management, the IPCC (1995) has drawn a number of conclusions.

Water management is a complex field; changes in one sector, such as water supply, will inevitably result in changes in other sectors, such as water consumption, ecosystems, etc. It is also evident that changes in water availability will affect many aspects of social and economic life, and these need to be taken into account in the consideration of other problems discussed elsewhere in the IPCC report. Most studies so far have considered future changes and have drawn conclusions without taking into account either the adaptive measures that are likely to be introduced, or the trends in other, non-climate factors that affect water resources. This is a rather simplistic and naive approach, because in practice a variety of adaptive water management measures are likely to be introduced in response to climate change.

In some cases non-climate factors could fully compensate for the effects of climate change, whereas in others they may intensify them. In some developed countries, for example, the level of water consumption per capita is falling as a result of more efficient water use, thus reducing vulnerability to changes in water supplies. In most developing countries, however, the opposite situation can be observed: their growing populations and higher levels of water demand will increase consumption, and thus their vulnerability to changes in climate and hydrological characteristics.

Under conditions of variable climate and increased human activity a number of opportunities for the introduction of adaptive measures are available. New scenarios are therefore needed to examine alternative long-term development strategies, taking into account different combinations of possible climate change factors, water resources control measures, and the ecological consequences. These strategies should be compared and estimated in relation to the level of reliability of water supplies, cost, and the environmental and socio-economic consequences. For example, some strategies may be better adapted to the uncertainty of climate change, thus improving the reliability of water supplies, whereas other strategies may focus on environmental protection, etc.

In generalizing these case studies, the IPCC (1995) concludes that future water management strategies should include cost-effective combinations of the following measures:

- direct measures to improve the control of water use (regulation, technological improvements);
- institutional changes to improve water resources control;
- improvement of water management systems operation;
- direct measures to improve water supplies (reservoirs, pipelines, etc);
- measures to improve water distribution technologies and the efficiency of water use.

Different strategies will of course be required for different conditions. For example, watersheds without runoff control will require quite different water management strategies compared with river systems with reservoir runoff control, including canals, dams, etc. Similarly, different adaptation strategies will be needed in expanding urban areas than in rural areas. As long as rational water management strategies are developed for individual regions, designed to meet all future local demands and requirements under a stationary climate situation, they are likely to be successful in preventing or mitigating many of the negative impacts of climate change.

2.7 THE IPCC'S CONCLUSIONS AND RECOMMENDATIONS

In its general conclusions, the IPCC has made the following recommendations and proposals for future activities:

- Scenarios based on GCMs and on paleoclimate reconstructions do not provide sufficiently reliable information for multipurpose assessments of the hydrological consequences of climate change at the scale of river basins.
- River systems are highly sensitive to climate change, although recent changes due to human interference cannot be distinguished from the wide variety of natural changes in hydrometeorological characteristics.
- The hydrological regimes of arid and semi-arid regions are highly sensitive even to slight anthropogenic changes in climate, particularly precipitation.
- In river basins where the major proportion of runoff is derived from snowmelt in mountainous areas, global warming will have a greater effect on the streamflow distribution throughout the year than on annual runoff. Variations in air temperature may have a more significant effect than changes in precipitation.
- Reliable quantitative conclusions on possible changes in extreme hydrological parameters cannot be drawn, but it is reasonable to assume that at the scale of river basins (small and medium-sized basins in particular), global warming will have a greater impact on extreme parameters than on annual and seasonal values.
- The direct effects of increased CO_2 concentrations in the atmosphere vegetation could be to reduce water uptake, lower transpiration and increased runoff.
- The vulnerability of water systems to climate change could be reduced by controlling runoff and improving the management of water resources.
- Changes in mean runoff parameters mean that more systematic and precise criteria will be required for water engineering projects, particularly in the operational procedures, optimization systems and contingency planning for existing and future water management systems.
- Existing systems of water resources control should be adapted to meet the increasing demand for water under conditions of climate change.
- Control of water demand and institutional adaptation are basic requirements for improving the stability of water resources systems under conditions of increasing uncertainty.
- Technological innovation and improved water management could help to mitigate many of the negative consequences of global warming.

2.8 COMMENTS ON THE IPCC'S ACTIVITIES

Over the last seven years the IPCC's work to assess the hydrological consequences of anthropogenic changes in the global climate has been useful and fruitful. Its success has been due to the Panel's commitments:

- to invite numerous specialists from both developed and developing countries to assess the status of this global problem;
- to stimulate research and studies of trends, as well as to improve our understanding of hydrometeorological processes;
- to compile and summarize the results of research – both published and unpublished – and information from around the world;
- to formulate scientific conclusions, and to develop recommendations and proposals for the future;
- to disseminate the obtained results among a wide circle of scientists, international organizations, decision-makers and the public;
- to strengthen contacts and improve cooperation among scientists engaged in the fields of hydrometeorology, ecology, economics and environmental protection.

Nevertheless, there have been some gaps and shortcomings in the IPCC's work. For example, it has not been possible to develop unified scenarios of the future climate nor to recommend on a selection of methods for assessing the hydrological consequences of global climate change.

Not all countries and physiographic regions are adequately represented in the IPCC's reports; for example, there is little information on the permafrost and mountain regions, and studies of large countries such as Brazil, China, India and Russia have so far been inadequate. Meanwhile, most of the references, particularly in the field of water resources management (IPCC, 1995) focus on the physical and socio-economic conditions of the USA. Few data are yet available for analyses of the actual effects of global warming on hydrological regimes in different world regions during the 1980s and 1990s.

For some regions the results of assessments of the hydrological impacts of climate change using all the basic scenarios have been similar. Generalizations have not yet been drawn from these results, but they could provide useful information for the development of adaptive measures.

Over the last five or six years the IPCC has made little progress in predicting the effects of global climate change with any degree of certainty, nor in improving our understanding of the processes of change in relation to environmental and socio-economic systems. In the future, multidisciplinary investigations of this global problem will be urgently needed.

3 Assessment of the impacts of climate variability and change on the hydrology of South America

B. P. F. BRAGA and L. C. B. MOLION

3.1 INTRODUCTION

Very few studies of the impacts of climate variability and change on water resources systems have been reported in the literature. Notably, the only published report dealing specifically with South America presents the efforts of the US Environmental Protection Agency in cooperation with the Federal University of Rio Grande do Sul, Brazil, to analyze the impacts of climate change in the Uruguay River basin. Another report (Stakhiv *et al.*, 1993) describes the results of an impact study in the Orinoco River basin in Venezuela. Unfortunately, the results of these studies were not available at the time of preparation of this report, nor those of a Russian–Argentinean effort in Argentina.

Due to its strategic importance for the world's climate, the Amazon basin has been the subject of many studies of hydroclimatology, involving Brazilian and other international institutions. These early efforts included the ARME and ABRACOS projects, developed by institutions in Brazil and the UK, whose objective was to assess the forest water consumption and exchange with the surrounding atmosphere. A major scientific experiment involving all the Amazonian countries and the international community is now being planned: the LAMBADA project (now part of the broader LBA project) will focus on the hydroclimatology of the region and will provide important data for future studies of the impacts of climate change on tropical regions.

This chapter describes the above studies and the planned experiment, and presents their main results. A common conclusion from all these efforts is the limited ability of general circulation models (GCMs) to describe accurately the convective precipitation process. In particular, soil–water–plant interactions with the atmosphere in the tropical environment are poorly simulated in the models.

3.2 CLIMATE VARIABILITY: SHORT- AND LONG-TERM TRENDS

The physical causes of the interannual variability of rainfall in South America are not yet adequately understood, but they are certainly linked with the spatial and temporal variability of the incoming solar radiation and the power of equatorial heat sources. One set of phenomena, known as the El Niño–Southern Oscillation (ENSO), greatly modifies both the positioning and the intensity of heat sources. During an El Niño event – characterized by higher than normal sea surface temperatures (SSTs) in the eastern equatorial Pacific and a negative phase of the SO – rainfall over the continental tropics is greatly reduced. This occurs because the ascending branch of the Walker circulation (the east–west global circulation component), which is normally located over the Amazon basin, is displaced westward over the abnormally warm equatorial Pacific, and is intensified due to strong convection. The descending branch of the Walker circulation covers the whole of northern South America and the adjacent Atlantic, reaching as far as Africa. The subsidence suppresses convection and rainfall is dramatically reduced. Figure 3.1 (Kayano and Moura, 1983) shows isolines of rainfall departures from the mean normalized by the standard deviation, during a particularly strong ENSO event in February–May 1958. Over northern South America and Africa there are noticeable reductions of over 1.8 times the standard deviation. During the strongest ENSO event of the century, in January–May 1983, some climate stations in Amazonia recorded rainfall reductions of over 30%, and in parts of northeastern Brazil the reductions were in excess of 80% (Kousky *et al.*, 1984). In contrast, southern and southeastern Brazil, northern Argentina, Uruguay and Paraguay experience excess rainfall and floods in El Niño years.

During a La Niña event – lower than normal SSTs in the Pacific and a positive phase of the SO – the behaviour is exactly the opposite; northern and northeastern South

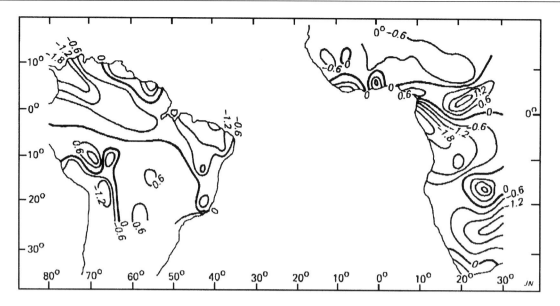

Figure 3.1 Isolines of rainfall deviations from the mean normalized by the standard deviation of the series for the period February–May 1958 (Kayano and Moura, 1983).

America receive excess rainfall, while droughts spread over southern and southeastern Brazil, northern Argentina, Uruguay and Paraguay. Figure 3.2 shows isolines of correlation coefficients between the Southern Oscillation Index (SOI) and rainfall over Brazil in September–November using the records of 41 years (Rao and Hada, 1990). Positive values in excess of 0.6 are found over northeastern Amazonia (56° W), indicating droughts (floods) during El Niño (La Niña) years. Over southern Brazil, the values are negative exceeding 0.7 and the rainfall behaviour is the opposite. Analyses of point rainfall data reflect neither the extent nor the magnitude of the impact of ENSO events because of the inadequate spatial distribution of rain gauges.

Molion and Moraes (1987) correlated a time series of SOI with the corresponding time series of discharges for selected rivers of South America. For the Trombetas River, a tributary of the Amazon near 56° W, the correlation coefficients were positive and higher than 0.8, implying that when the SOI is negative (El Niño years) discharges are below normal, and when it is positive (La Niña years) discharges are above normal. The authors also noted that there is a three-month time lag between the SOI and the Trombetas discharges, and suggested that the SOI, being the leader, might be used to forecast river discharges. For the Paraná River, on the other hand, the correlation coefficients were small but negative, suggesting that discharges are above (below) normal during El Niño (La Niña) years.

In the southern hemisphere cold fronts or frontal systems form an important dynamic mechanism for the occurrence of rainfall in northern South America. In some years, however,

blocking patterns are established, preventing the penetration of frontal systems from lower latitudes. In an analysis of 90 years of rainfall data, Molion (1994) showed the coincidence of years with below normal rainfall and the presence of volcanic aerosols in the stratosphere over low latitudes. Such aerosols can reduce the incoming solar radiation, sometimes by more than $10 \, W/m^2$, as observed during the eruption of Mount Pinatubo (Minnis *et al.*, 1993). They cool the tropical troposphere over the continent, weakening the equator to pole temperature gradient, and reduce evaporation from the Atlantic Ocean, the main source of moisture for the South American continent. In the cooler and drier, and thus more stable troposphere, convection is inhibited. The upper troposphere divergence and the Hadley circulation cell weaken. Atmospheric pressure increases, thus weakening the ocean–continent pressure gradient and, therefore, the trade winds. The subtropical anticyclone appears to extend its influence over the continent and is displaced poleward, blocking the northward incursion of the South American Convergence Zone (SACZ); that is, the anticyclone system blocks the penetration of frontal systems and the southward displacement of the Intertropical Convergence Zone (ITCZ). This is depicted schematically in Figure 3.3. The result is the occurrence of droughts in the Amazon and Orinoco basins and northeastern Brazil. Conversely, there is excess rainfall in the central and southeastern parts of the continent, mainly over the Paraná–La Plata basin and the southeastern coast.

Long-term or low-frequency climate variability has been observed over both the Amazon and Paraná River basins. From the late nineteenth century to the 1920s, the continen-

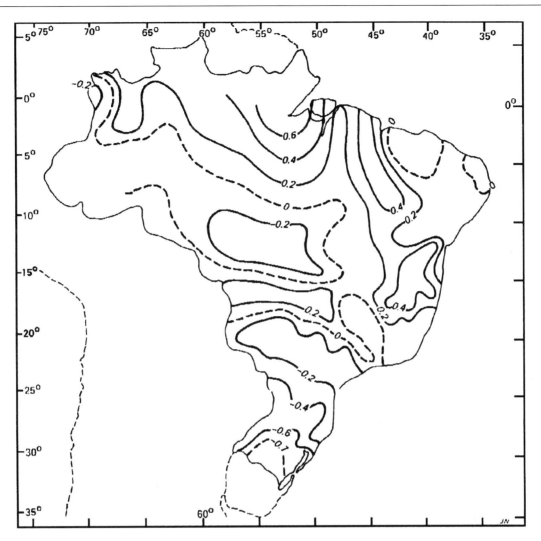

Figure 3.2 Isolines of correlation coefficients between the Southern Oscillation Index and rainfall for Brazil for the trimester September–November (Rao and Hada, 1990).

tal tropics were reported to have been generally drier than at present. One of the causes might have been the higher frequency of ENSO events, whose occurrence is known to be associated with below normal rainfall in the tropics (Molion, 1990). Between 1851 and 1930 there were nine strong and 16 moderate ENSO events, averaging one event per 3.2 years, whereas between 1931 and 1990, there were only five strong and seven moderate events, i.e. one event per 5 years (data from Quinn and Neal, 1987). For example, in the period 1911–30, Manaus, in the middle of Amazon rainforest, had an average of 153.9 ± 19.7 rainy days per year, whereas in the period 1931–60 this average had increased to 189.3 ± 15.0 rainy days per year. Even more important, during the dry season (July–September) there was also a 34% increase in the number of rainy days, from 19.6 ± 6.2 in 1911–30, compared with 26.3 ± 7.0 rainy days in 1961–90.

The dryness of the beginning of the century was also apparent in the level of the Negro River, measured at Manaus, which reflects the behaviour of that large forested basin. Between 1903 and 1926, six of the Negro River's maximum annual (flood) levels ranked amongst the lowest ten maxima; the flood level in 1926 was more than 5 standard deviations below the 90-year mean – the lowest ever recorded.

Barros *et al.* (1995) analysed long-term rainfall data recorded at about 70 climatological stations between 22°–53° S and 48°–72° W, and found a positive rainfall trend of 9 mm/year during the period 1956–91 in the Paraná River basin (Figures 3.4 and 3.5). They attributed the observed 35-year trend to the decrease in the mean meridional temperature gradient, which diminished by 1.5°C during that period. One of the consequences was a 5° latitudinal southward displacement of the global circulation systems, particularly the

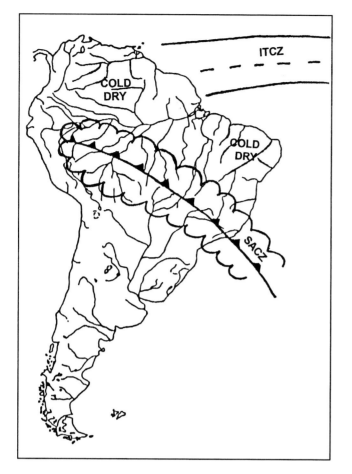

Figure 3.3 The effects of volcanic aerosols. The relatively drier and colder tropical troposphere blocks the northward displacement of the SACZ and the southward displacement of the ITCZ (Molion, 1994).

South Atlantic subtropical anticyclone, which favours the inflow of humid air and causes cold fronts to remain stationary over the region.

Molion (1995) noted the apparent coincidence of rainfall variability over the whole continent with long-term volcanic activity. During the period 1915–55 there were no major volcanic eruptions and the optical depth of the atmosphere decreased to the lowest values reported in the past 100 years. The subsequent increase in incoming solar radiation reaching the surface lowered the atmospheric pressure over the tropics and strengthened the mean meridional pressure gradient, forcing the more frequent penetration of frontal systems into lower latitudes. This enhanced convection and rainfall over Amazonia and northeastern Brazil. At the same time, the rainfall over the Paraná River basin was apparently unaffected (Figure 3.4). The volcanic activity resumed in 1956, cooling and drying the tropical troposphere, weakening the mean meridional temperature gradient and enhancing rainfall over the Paraná basin (Barros *et al.*, 1995). Over Amazonia,

however, there were no significant changes in rainfall totals in 1915–55, perhaps due to the fact that the world oceans also warmed. Since the heat capacity of the oceans is greater than that of continental areas, appreciable long-term climate variability in the tropics might lag by two to three decades. That is, tropical convection, and therefore rainfall, may still be fed by moisture coming from warmer than normal oceans.

Global warming due to human activities may also modify the South American climate. Burgos *et al.* (1991) attempted to assess the possible changes by analysing the changes in some climatological and dynamic atmospheric features under a scenario for the years 2010 and 2050, taken from Hansen *et al.* (1988). In essence, global warming may impose on South America a climate similar to that of the period 1915–55 when volcanic activity was the lowest over the past 200 years.

3.3 IMPACTS OF CLIMATE CHANGE ON THE HYDROLOGY OF SOUTH AMERICA

3.3.1 The US EPA–University of Rio Grande do Sul experiment

This experiment was conducted by the University of Alburn, USA, and the Federal University of Rio Grande do Sul, Brazil, under the auspices of the US Environmental Protection Agency (Tucci and Damiani, 1991). The EPA's objective was to study major river basins in different continents. In South America, the Uruguay River, a tributary of the La Plata, was chosen because of the availability and reliability of hydrological data, its extensive agriculture, the existence of a number of urban centres, and the relatively large number of planned hydropower plants. The water resources of the Uruguay River are shared by Brazil, Argentina and Uruguay.

The La Plata River basin has three main rivers, the Paraguay, the Uruguay and the Paraná, which drain an area of about 3.1 million km^2 (Figure 3.6). The precipitation distribution in this basin decreases upstream to downstream and increases from west to east. The total annual rainfall varies from 100 mm close to the Andes, up to 2000 mm near Serra do Mar (the eastern coastal range divide). The left banks of the Paraná and Uruguay Rivers receive more than 1200 mm rainfall per year, while the opposite bank of the Paraná River and the middle and lower regions of the Paraguay basin receive less than 1000 mm per year. The upper La Plata basin has a tropical climate, which changes to temperate and then to subtropical as one moves in the downstream direction.

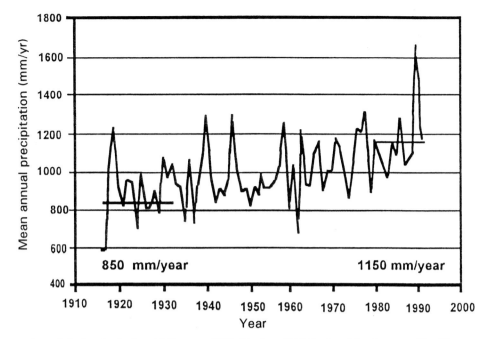

Figure 3.4 Mean annual precipitation in the humid Pampas, 1910–90; since 1950 the trend has been positive (Barros *et al.*, 1995).

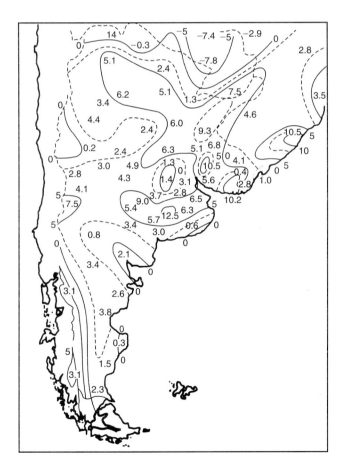

Figure 3.5 Rainfall linear trends (mm/year), 1956–91. Note that values exceeding 9 mm/year were recorded over the Paraná and La Plata basins (Barros *et al.*, 1995).

The Uruguay River (Figure 3.6) arises in the eastern coastal range of Brazil, the Serra do Mar, which forms the border between the states of Santa Catarina and Rio Grande do Sul. Further downstream it forms the border between Brazil and Argentina, and in its lower reaches between Argentina and Uruguay. The drainage area, estimated to be 276 000 km^2, can be divided into two major contributing areas. The upper Uruguay is 606 km long with a mean slope of 0.09 m/km, and the lower Uruguay is 348 km long with a mean slope of 0.03 m/km. Precipitation over the whole basin is characterized by a bimodal distribution.

3.3.2 Results of GCM studies

Three general circulation models (GCMs) were utilized in the experiment: the GISS (Goddard Institute for Space Studies), the GFDL (Geophysical Fluid Dynamics Laboratory) and the UKMO (UK Meteorological Office) models. In the discretization scheme of the GISS and GFDL models, the whole of the Uruguay River basin is represented in a single pixel, whereas the UKMO model produces two pixels of information, one each for the upper (UKMO1) and lower (UKMO2) Uruguay basins. The three GCMs were each run with two scenarios: scenario 1, assuming CO_2 concentrations at the level of the 1960s, or approximately today's levels; and scenario 2, assuming 'business-as-usual' conditions, leading to a doubling of CO_2 by the year 2060.

BASIN	AREA (Km²)
1 - RIO BONITO	
2 - PONTE ALTA DO SUL	4600
3 - PASSO MAROMBAS	3800
4 - SANTO ÂNGELO	5414
5 - POTIRIBU	628
6 - CONCEIÇÃO	784
7 - VILA CLARA	2783
8 - PASSO DA CACHOEIRA	2562

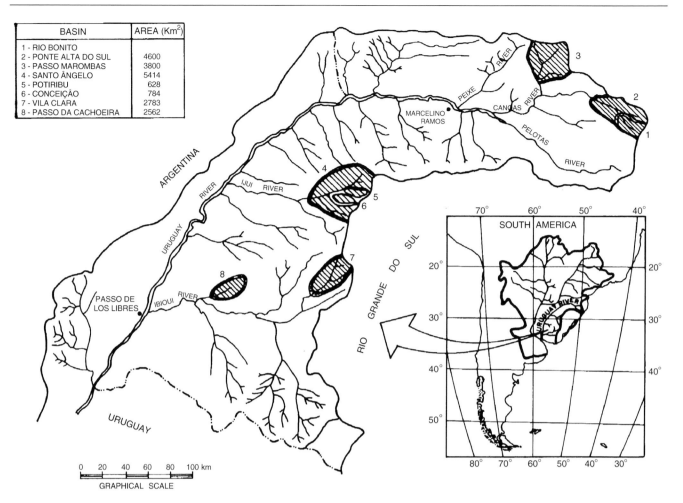

Figure 3.6 The Uruguay River basin.

The results were monthly averages of temperature and rainfall for each pixel considered for the two scenarios. The model results for scenario 1 were then compared with data collected from ten hydrometeorological stations located within the pixel, averaged over a period of 20 years. Figure 3.7 shows the results for temperature, and Figure 3.8 those for precipitation; it can be seen that the GISS model presented the most consistent results, in terms of both temperature and rainfall. However, as in other applications of GCMs for other parts of the world, all of the models significantly overestimated the monthly rainfall totals.

The GCMs were then used to compute monthly rainfall and temperature variations under the $2 \times CO_2$ scenario, while evapotranspiration was computed by the Penman and Thornthwaite methods based on predicted temperatures for the year 2060. Figures 3.9 and 3.10 show the results for temperature and precipitation. Although all of the models showed increases in temperature, the amounts of the increases differ significantly: the GFDL model showed an average of $+ 3.6°C$, whereas the UKMO2 model showed

$+ 6°C$. The GFDL results showed a significant increase in rainfall in October ($+ 154\%$) and a large decrease in December ($- 46\%$), with an increase in average annual precipitation of 19%, whereas UKMO1 also showed an increase in annual precipitation of 19%. These results conflict with those of the UKMO2 model, which showed an increase in precipitation of only 4%, and of the GISS model, which predicted a 2% decrease, with the largest decreases in April and May (17 and 14%, respectively).

These monthly increases/decreases in precipitation were then applied to existing records of rainfall, temperature and evapotranspiration to obtain a new series of input data for the rainfall/runoff model, which in turn produced streamflow forecasts under this scenario.

3.3.3 Impacts on streamflow and water use

The IPH-III hydrological model (Mota and Tucci, 1984) was used for transforming rainfall into runoff. This model uses a loss function for interception and evaporation. A modified

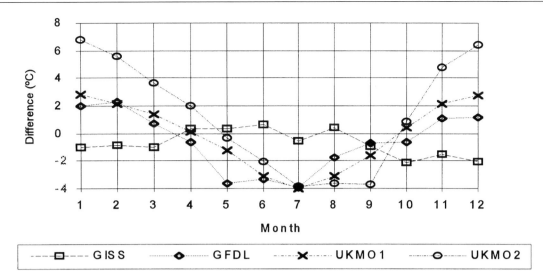

Figure 3.7 Difference between recorded and simulated temperatures by various GCMs.

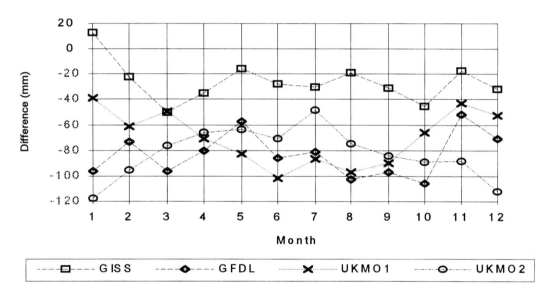

Figure 3.8 Difference between recorded and simulated precipitation by various GCMs.

Horton mechanism was used to separate surface runoff volumes from infiltration. Overland flow routing was performed using the Clark isochrone method, and groundwater flow routing was modelled using a simple linear reservoir. Streamflow routing uses a kinematic wave approximation. The model operates on a daily basis and was applied to each of the eight river basins indicated in Figure 3.6. Several criteria were used for the calibration, including statistics such as the coefficient of determination and minimization of peak and base flow differences. The model was verified using a record of daily flows not used in the calibration procedure. Calibration showed R^2 in the range of 0.85, and in verification tests this value fell to an average of 0.76. For both calibration and verification, 20 years of daily rain-

fall and monthly potential evapotranspiration data were used, with Penman's equation.

The impact of climate change was simulated using the monthly precipitation totals multiplied by the ratio of the precipitation forecast under the $2 \times CO_2$ scenario and the present monthly precipitation. Future monthly temperature values were obtained by adding the expected monthly temperature increase to the present monthly temperatures. Evapotranspiration for the future scenario was obtained using Penman's equation with the forecast future temperatures alone.

Since the hydrological model operates on a daily basis, the monthly precipitation values had to be disaggregated into daily values, using a historical disaggregation pattern. This

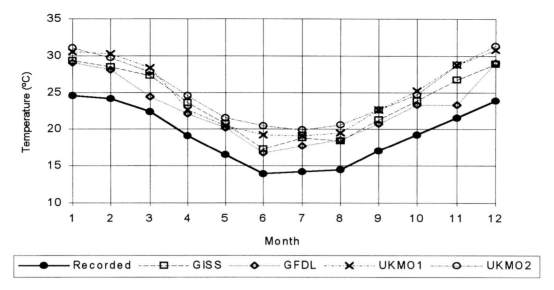

Figure 3.9 Temperature variations under the $2 \times CO_2$ scenario.

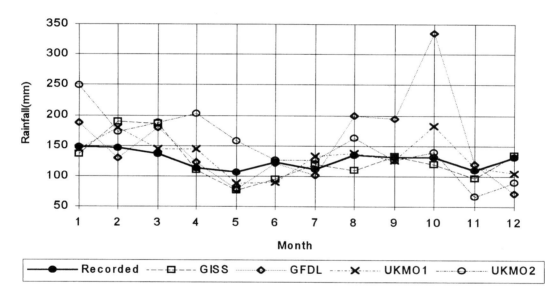

Figure 3.10 Precipitation variations under the $2 \times CO_2$ scenario.

implies that the temporal distribution of rainfall will not be affected by climate change. The results in terms of the mean daily flows for the eight basins are presented in Table 3.1 (Tucci and Damiani, 1994). The results of the three models are contradictory: the GFDL model shows that a doubling of CO_2 would cause an average increase in runoff of approximately 20%, whereas the GISS model suggests an average reduction of the order of 10%. More surprising still are the results of the UKMO model, which suggest increases in runoff in some basins, and decreases in others separated just a few hundred kilometres.

To obtain a better impression of the temporal variability of the runoff, Tucci and Damiani (1994) plotted the mean

monthly specific flows of the eight basins. The results are shown in Figures 3.11–3.14. The GISS model shows flow reductions from April to January, which would shift the critical drought period to October–December. The GFDL model results were affected significantly by the large rainfall forecast for October. The UKMO1 model showed a variable pattern of flows throughout the year, while UKMO2 showed a tendency for flows to increase for most of the year.

The same study also analysed the impact of climate change on the main uses of water in the Uruguay basin: hydropower, flood control, water supply, irrigation and wastewater disposal. Hydropower is an important use of water in the basin; the potential capacity is 16 500 MW, of which only 2680 MW

Table 3.1. *Percentage variations in mean daily flows in the eight basins, under the $2 \times CO_2$ scenario (Tucci and Damiani, 1994).*

Basin	GISS	GFDL	UKMO
Potiribu	− 13.7	15.6	10.8
Conceição	− 11.1	13.9	4.8
Santo Angelo	− 12.2	20.0	9.5
Ponte Alta do Sul	− 9.62	23.6	− 2.6
Bonito	− 10.4	24.9	− 2.6
Marombas	− 10.4	32.5	− 2.4
Passo da Cachoeira	− 12.4	16.4	12.5
Villa Clara	− 14.1	25.2	21.2

have been installed. The analysis was performed considering all existing and planned hydropower plants in the basin. In addition, a theoretical analysis was conducted to estimate the impacts of climate change on small hydropower plants that may be built in the future. According to the GISS model, a 5% reduction in hydropower production can be expected, whereas the UKMO model indicated a reduction of 2.5%, and the GFDL an increase of 17.3%. Small hydropower plants can be expected to be negatively impacted according to all GCMs.

Floods are expected to cause fewer problems in already critical areas (Uruguaiana, São Borja, Itaqui and Marcelino Ramos). With the exception of the GFDL model, which showed increases in flood depths of nearly 4 m, the other two models showed reductions of 0.75–1.10 m, depending on the return period considered. Water supplies are likely to be doubly impacted by climate change: on the demand side with an increase in temperature, and on the supply side with a decrease in precipitation. The former factor was not taken into account in this analysis, but a typical 20 km^2 watershed was used to analyse the impact of reduced precipitation. The available water supplies are likely to be reduced by about 10%, even with the use of regulating reservoirs.

3.4 INTERNATIONAL EXPERIMENTS RELATED TO CLIMATE IN SOUTH AMERICA

Since 1983 the Instituto Nacional de Pesquisas Espacias (INPE, Brazilian Space Research Institute), Instituto Nacional de Pesquisas da Amazônia (INPA, Brazilian Institute for Amazon Research) and the Institute of Hydrology in Wallingford, UK, have been conducting research in the Brazilian Amazon related to hydrology, meteorology and hydroclimatology (Molion, 1987). This has resulted in a series of interconnected experiments that have allowed some scientific speculations about the consequences of Amazonian deforestation for the global climate using GCMs (e.g. Shukla *et al.*, 1990; Lean and Warrilow, 1989). Other international institutions in the USA and Europe have collaborated with Brazilian institutions to analyse hydrological and meteorological processes in the Amazon basin. The most significant efforts in the region are described in the following.

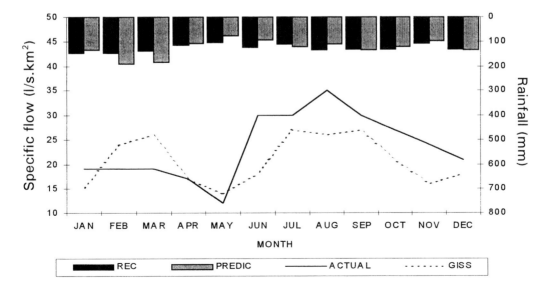

Figure 3.11 Mean monthly specific flow hydrographs produced by the GISS model for the Uruguay River basin.

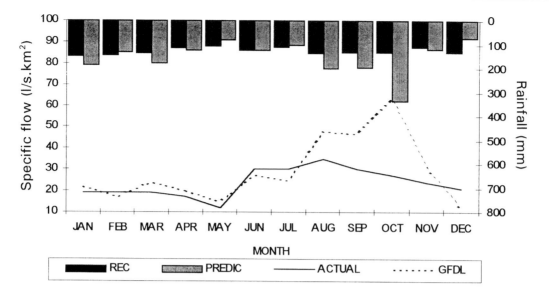

Figure 3.12 Mean monthly specific flow hydrographs produced by the GFDL model for the Uruguay River basin.

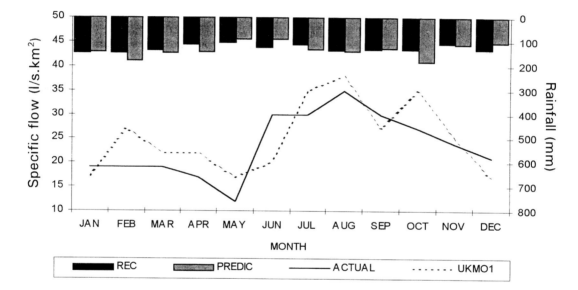

Figure 3.13 Mean monthly specific flow hydrographs produced by the UKMO1 model for the Uruguay River basin.

3.4.1 The Amazon Region Micrometeorological Experiment (ARME)

The Amazon Region Micrometeorological Experiment (ARME) was initiated in September 1983 with a two-year time span to study surface–atmosphere water energy exchanges. A single scaffolding micrometeorological tower, 45 m high, reaching 10 m above the forest canopy, was installed 25 km northwest of Manaus in central Amazonia. Continuous measurements were made of the following variables: precipitation, interception, evaporation from the canopy, soil moisture and ground heat flux (Molion, 1987).

Precipitation throughfall was measured using 100 rain gauges on the ground beneath the canopy. In addition, routine automatic measurements, and four intensive field expeditions of two to three months were undertaken to collect information related to the surface energy exchange, aerodynamic exchange characteristics of the forest, and to record vertical air temperature, humidity and wind speed gradients (Shuttleworth, 1989). Other studies focused on plant physiology. These data have been used extensively in the development of tropical vegetation interaction schemes for use in climate models.

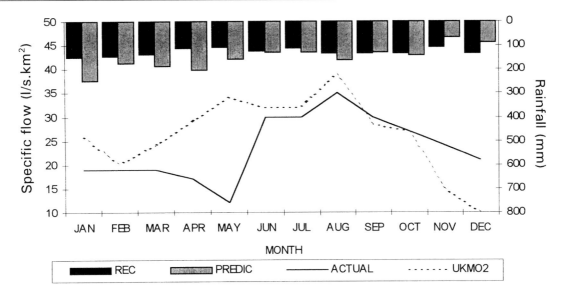

Figure 3.14 Mean monthly specific flow hydrographs produced by the UKMO2 model for the Uruguay River basin.

The ARME data showed that the central Amazonian rainforest has a shortwave albedo of 12–13%. On average, 50% of the annual rainfall is evaporated, consuming in the process 80–85% of the net radiation. Rainfall intercepted by the canopy is 10–15% of the total, and corresponds to 25% of the annual evaporation. The remaining 75% of the total evapotranspiration occurs as transpiration, assuming no significant restrictions on the availability of soil water (Molion, 1987; Nobre and Shuttleworth, 1993). The data obtained by this project were used by Sellers *et al.* (1989), Sato *et al.* (1989) and Nobre *et al.* (1991) to improve the surface parameters in GCMs applied to assess the impact of deforestation on the Amazonian climate. The results suggest regional reductions in evapotranspiration (by $20 \pm 10\%$) and rainfall (by $30 \pm 20\%$), and an increase in temperature of $2 \pm 1\,\mathrm{K}$. The reduced rainfall leads to a 10% reduction in runoff.

Other independent studies of the impact of deforestation on precipitation in Amazonia using GCMs have produced similar results. Lean and Warrilow (1989), for example, assumed that throughout the entire Amazon basin the tropical rainforest would be replaced by a savanna biome. They adopted a simple land surface treatment, including the assumption of the same temperature for the soil and the forest canopy. In a subsequent study, Lean and Rowntree (1993) improved the early version of the model by including a description of the process of evaporation of intercepted rainfall. Although the Lean and Warrilow experiment showed a 20% decrease in rainfall, Lean and Rowntree obtained a slightly lower decrease of 14%. Later, Dickinson and Kennedy (1992), using a modified version of the NCAR GCM1 model coupled with the biosphere–atmosphere transfer scheme (BATS; Dickinson, 1984),

obtained even larger reductions in precipitation in a three-year simulation of deforestation of the Amazon basin.

Although all of these exercises with GCMs represent important scientific attempts to analyse the impacts of deforestation on climate in the Amazon basin, it is perhaps opportune to quote Dickinson and Rowntree (1993) in this regard: 'In general, the models suffered from a lack of realism in the spatial and temporal distribution of precipitation, excessive incident solar radiation, and an overly large fraction of evapotranspiration originating from canopy interception. While some of these model defects are becoming less serious, the role of inter-annual variability in tropical rainfall has yet to be addressed and the subject is still too exploratory to allow the development of future scenarios.'

3.4.2 The Anglo–Brazilian Amazonian climate observation study (ABRACOS)

The ABRACOS experiment started in 1990 with a four-year time span to study the interactions between vegetation cover and the atmosphere in relation to the water cycle and energy fluxes. The project was sponsored by the UK Overseas Development Agency and the Brazilian government, and involved Brazilian research institutions (INPE, INPA and CENA) and the Institute of Hydrology in the UK.

The experiment involved data collection from three locations, in areas with different ecological characteristics. Thus valuable data on aerodynamic roughness is now available for rainforest (1.5 m) and pasture (0.2 m). It was observed that during dry periods, there are distinct differences in the diurnal cycles of surface temperature and thermal energy exchange between pasture and forest vegetation (Nobre and Shuttleworth, 1993). In addition, preliminary results show that areas of pasture are more sensitive to rainfall variations since grasses extract water for evapotranspiration from only the first 1 m of soil, whereas rainforest trees can extract water from as deep as 8 m. These results will allow scientists to improve their models of the land surface as well as those of soil moisture exchange processes. The Hadley Centre for Climate Prediction of the UK Meteorological Office and the Centro de Previsão de Tempo e Estudos Climaticos (CPTEC) in Brazil are now using the new dataset in their GCMs (Lean *et al.*, 1996).

3.4.3 Carbon in the Amazon River experiment: CAMREX

The University of Washington, INPA and the Centro de Energia Nuclear na Agricultura (CENA, National Center Nuclear Energy in Agriculture) have been conducting a long-term, large-scale experiment in the Amazon basin to assess the biogeochemical dynamics of the entire basin. The research objectives include quantification of the spatial and temporal distribution of water, dissolved and particulate organic matter and related bioactive elements throughout the basin (Victoria *et al.*, 1993). The project also aims to provide runoff discharge data for large-scale studies related to the impacts of climate variability and change in the Amazon basin. An important aspect of CAMREX is that it focuses on the relationship between geomorphology, soils and land use, and near-surface water and sediment budgets.

3.4.4 The Amazonian boundary layer experiment (ABLE-2) and the Amazon moisture flux experiment (FLUAMAZ)

Two large-scale experiments were begun in July–August 1985 (ABLE-2) and April–May 1987 (FLUAMAZ) to explore the atmospheric chemistry of the Amazon region. The experiments were an initiative of the Atmospheric Chemistry Program of NASA, together with the US National Science Foundation, the National Center for Atmospheric Research (NCAR) and INPE. In Brazil, field support was provided by INPE and INPA. The work was conducted at a site near Manaus. Aircraft, satellite platforms and ground-based equipment were used to measure the dynamics of the lower atmosphere over the entire Amazon rainforest in wet and dry conditions.

The FLUAMAZ experiment was designed to determine the sources, sinks, concentrations and transport of trace gases and aerosols originating from tropical rainforest soils, wetlands and vegetation. A number of small-scale meteorological measurements have been made, and basin-scale moisture budgets have been derived. A network of six radio-sounding stations has tracked soundwaves that were released four times per day for a month. An important result of this experiment is that evapotranspiration accounts for about 55% of the precipitation over the basin, corroborating previous results by Molion (1976) and Salati *et al.* (1978).

3.4.5 Hydrology of tropical ecosystems: HYTRECS

Hydrology plays a fundamental role in explaining the consequences of human activities on ecosystems in the humid tropics. Changes in the water cycle and sediment budgets can be properly analysed only with a clear understanding of hydrological processes on a watershed basis. However, traditional methods of hydrological analysis developed for temperate conditions need to be used with caution in tropical environments, since the hydrological and meteorological conditions are significantly different.

Preliminary estimates of the Amazon water balance have shown that nearly 50% of the rainfall in the basin is recycled from evapotranspiration (Molion, 1976). Deforestation will have significant impacts on different phases of the water cycle in tropical ecosystems. Another important issue is the sediment yield from deforested land. Any attempt to develop humid tropical areas on a sustainable basis must be preceded by watershed modelling; future models should be physically based to allow extrapolation outside the range of calibration.

An associated and more complex problem is the impact of deforestation on the local and global climate. Because of its size and position on the equator, the Amazon basin is a major heat source driving the general circulation of the atmosphere. Changes in land cover will alter the processes of exchange of sensible and latent heat, and their momentum, between the surface and the adjacent atmosphere. In the future, large-scale changes in land cover in the Amazon could change the general circulation of the atmosphere, promoting climatic change on a global scale (Molion, 1993).

In 1992 UNESCO and UNEP signed an agreement to analyse the hydrology of tropical ecosystems and the impact of deforestation on the hydrological cycle. The aim of the research effort was to improve understanding of the hydrological and erosion processes in humid tropical ecosystems on a watershed basis. To this end, two small basins in the Amazon were selected, one containing natural rainforest and the other pasture land. These small basins have been monitored on a continuous basis since October 1992, and a physically based model of the basins is now being calibrated

(D'Angelo and Damazio, 1993). The basins are located in eastern Amazonia near the city of Maraba, in Para state. Rain and stream gauge recording instruments have already been installed on two tributaries of the Tocantins River, with different types of vegetation cover.

The rainfall, streamflow, soil moisture and evapotranspiration data generated by HYTRECS will be used in the LBA experiment (see below) to improve the land surface descriptions in GCMs to assess the impacts of climate change on the Amazon basin.

3.4.6 The LBA experiment

The Large-scale Biosphere–Atmosphere Experiment in Amazonia (LBA)* encompasses the entire Amazon basin. Within the LBA, the LAMBADA (Large-scale Atmospheric Moisture Balance of Amazonia using Data Assimilation) has been designed to gain an understanding of the regional-scale transport of heat, moisture and other biogeochemical cycles of important elements such as CO_2. BATERISTA (BiosphereAtmosphere Transfers and Ecological Research In Situ Studies in Amazonia) involves embedded meso- to microscale analyses of fluxes and cycles to provide detailed data for the calibration and validation of models developed within LAMBADA. These two experiments were set up by the WCRP/IGBP Joint Working Group on Land Surface Experiments, with contributions from many independent researchers from around the world and supported by ISLSCP, IGPO/GEWEX and UNESCO. The conceptual plan for the LBA experiments (Nobre *et al.*, 1993) established a timetable for the measurements, which will begin only in 1997–98 due to the heavy instrumentation required, including satellites, radar equipment and aircraft.

LAMBADA

The LAMBADA experiment will rely on large-scale observation techniques to monitor variables that will yield the atmospheric and soil water budgets, together with energy fluxes on a basin-wide basis. Basically, two equations provide the water budget (Nobre *et al.*, 1993):

$$\int_{z_s}^{z_t} \frac{dq}{dt} dz = \int_{z_s}^{z_t} \nabla.vq \quad dz + E - P, \tag{3.1}$$

$$\int_{h_b}^{h_s} \frac{dS}{dt} dh = P - E - R_0, \tag{3.2}$$

where z_s is the elevation of the land surface (m), z_t is the elevation of the top of the atmosphere (m), h_b is the elevation of the bottom of the unsaturated zone (m), h_s is the land

surface elevation (m), q is the water vapour concentration (l/m^3 air), v is the wind velocity vector (m/s), E is the evaporation rate (mm/s), P is the precipitation rate (mm/s), and R_0 is the total runoff rate (mm/s), and S is the soil moisture content (l water/m^3 soil). Figure 3.15 shows the areal arrangement of the data collection stations over the Amazon basin.

A complex monitoring system will provide spatial and temporal averages of precipitation, runoff, atmospheric water vapour and wind. Evapotranspiration will be computed as a residual, and over large time scales soil moisture changes will be neglected. A surface network of rain gauges and two meteorological radar instruments will provide ground precipitation data. The Geostationary Operational Environmental Satellite (GOES) and the Tropical Rainfall Measurement Mission (TRMM) satellite will allow extrapolation of rainfall information to a larger scale. Runoff will be estimated from stream gauge data collected at stations of the Brazilian National Department of Water and Power on the main tributaries of the Amazon.

BATERISTA

The BATERISTA experiment has been designed to obtain information on the exchanges of water, energy, and carbon and trace gases on a microscale basis. It is expected that BATERISTA will provide important data for ecophysiological modelling in tropical environments. The data collected at different locations in the basin will provide calibration information for the large-scale models to be developed under LAMBADA. The layout of the experiment, shown in Figure 3.16, consists of a main research site (Ji-Paraná) and four secondary areas. IGBP and UNESCO have recently provided additional hydrological support (Becker, personal communication, 1995), and suggestions with regard to data collection and mathematical modelling.

Within the main research site there will be at least five land cover types: virgin forest, freshly cleared and established pastures, and two areas of regrowth. This will allow an assessment of the influence of land cover on CO_2, water vapour and sensible heat fluxes. This arrangement will also provide information on how mesoscale atmospheric features may be determined by such land cover transitions and consequently may cause changes in the local climate.

---

* The LBA experiment, established in 1995, consists of the formerly independent experiments LAMBADA (Large-scale Atmospheric Moisture Balance of Amazonia using Data Assimilation). BATERISTA (Biosphere–Atmosphere Transfers and Ecological Research *In Situ* Studies in Amazonia), and AMBIACE (Amazon Ecology and Atmospheric Chemistry Experiment).

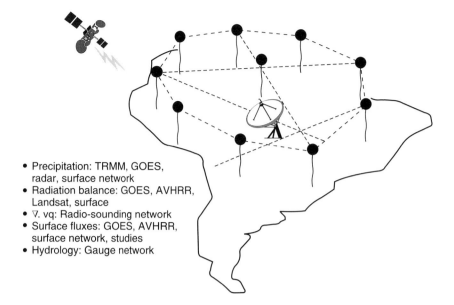

Figure 3.15 The LAMBADA experiment (Nobre *et al.*, 1993).

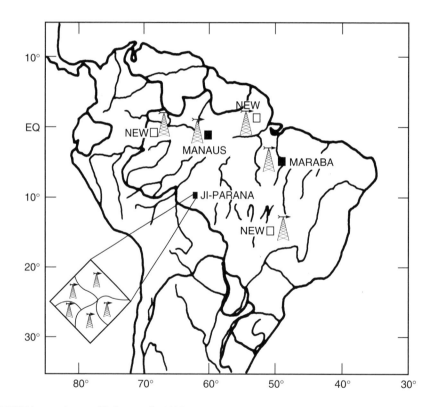

Figure 3.16 The BATERISTA experiment (Nobre *et al.*, 1993).

3.5 SUMMARY AND CONCLUSIONS

This chapter has described the initial major attempts to assess the impacts of climate change on the hydrology of some watersheds in South America. The main experiment described is the joint EPA–Federal University of Rio Grande do Sul project, in which three different GCMs were applied to the Uruguay River basin in the borders of Brazil, Uruguay and Argentina. The temperature and precipitation results of these GCMs were used in a rainfall–runoff model to produce monthly streamflow forecasts for eight sub-basins of the Uruguay River. The study also analysed the impacts of climate change on the main water uses in the basin. The results showed marked differences in the results of the GCMs; two predicted lower runoff and the third predicted an increase in monthly flows in the basin. From this important scientific effort it is clear that there is still a long way to go before GCMs produce reliable scenarios for future climate.

A series of experiments is now being conducted in the Amazon basin to obtain information that can be used to improve the quality of GCM forecasts for tropical regions. One of these is a UNESCO–UNEP initiative in which hydrometeorological and streamflow data are being collected from two experimental basins on a continuous basis. The LBA, a large-scale experiment involving several scientific institutions, is planned to start in 1997. It is hoped that within the LBA, the LAMBADA experiment will improve our understanding of the interactions and exchanges of water and energy fluxes in the world's largest tropical river basin, and that the associated BATERISTA experiment will allow a more refined analysis of hydrological processes in tropical ecosystems.

It is hoped that the results of these experiments become available to the scientific community, and can be used to provide better descriptions of the hydrological cycle in the soil–plant–atmosphere system that can be implemented within the structure of the existing GCMs.

4 Assessment of the impacts of climate variability and change on the hydrology of North America

G. H. LEAVESLEY

4.1 INTRODUCTION

Historical evidence of the effects of climate variability on the hydrology of North America is found in paleorecords interpreted from a variety of sources, including ice cores, lake sediment cores, tree rings, and pack rat middens. These records help identify the occurrence, duration, and effects of climate extremes that have ranged from cold and wet periods of major glaciations, to warmer and drier periods that were ice-free. Recorded climatological and hydrological data from the past few hundred years provide more detailed measures of the effects of climate variability, including extreme events such as the drought of the 1930s and significant floods such as the recent Upper Mississippi River floods of 1993.

The effects of climate variability and change on the hydrology and water resources of different regions of North America have been investigated by a variety of researchers for several decades. The most recent concerns regarding the effects of increasing greenhouse gases on climate variability and change have expanded these research efforts (1) to better define 'natural' climate variability; (2) to look for the effects of greenhouse gas increases on the climate and hydrological systems; and (3) to develop methodologies to estimate the effects of future climate variability and change on hydrological systems.

A variety of empirical, conceptual and physically based models have been employed using historical data and the estimates of future climate scenarios. Early investigations of climate change used hypothetical climate scenarios based on historical records and investigators' 'best estimates' of possible changes in temperature and precipitation. More recent studies have used climate scenarios developed from general circulation models (GCMs). Given the large degree of uncertainty in hypothetical and GCM-based scenarios, most of these studies can be viewed primarily as investigations of the sensitivity of water resources to a possible range of changes in temperature and precipitation.

Research efforts to date have identified a wide range of water resource sensitivities to climate variability and potential climate change. These studies have also identified a large number of gaps in our basic understanding of climatic and hydrological processes and their interactions. To address these knowledge gaps, a number of national and international research programmes have been established to facilitate and fund basic and applied research to investigate the coupled atmospheric and hydrological systems so as to provide more reliable estimates of potential climate changes and their impacts.

The objectives of this chapter are (1) to present an overview of the climate–hydrology relations of North America and the potential for changes in these relations as the result of climate change through a review of selected research results that explore these relations; and (2) to provide a brief overview of major national and international research programmes that are being initiated and/or conducted to address the gaps in our knowledge of climate–hydrology interactions and to develop the tools needed for objective assessments of the impacts of climate variability and change.

4.2 HYDROCLIMATOLOGICAL REGIMES AND THEIR VARIABILITY OVER TIME

North America has an area of about 21.5 million km^2 (Espenshade and Morrison, 1978) and consists of the countries of Canada (~10 million km^2), the United States (~9.5 million km^2), and Mexico (~2 million km^2). Climate ranges from tropical in parts of Mexico and the southern United States, to arctic in northern Canada and a large part of the US state of Alaska. The large latitudinal and elevational gradients found across North America strongly influence the variability of its climate in terms of the spatial and temporal distribution of hydrologically important variables such as precipitation, temperature, and solar radiation. A wide variety of hydrological regimes are produced across the continent as the result of the diversity in the magnitude and variance of these climate variables and their interactions with landscape variables, such as geology, soils and vegetation. The addition of anthropogenic factors to these interactions further increases the complexity and diversity of these hydrological regimes.

The climatology of North America has been described in detail by Barry and Chorley (1987). The major sources of moisture for precipitation over North America are the Pacific Ocean and the Gulf of Mexico. The Pacific Ocean is the primary source of moisture for areas west of the Rocky Mountains, while the Gulf of Mexico provides moisture to the central and eastern regions of the continent. The influence of the Atlantic Ocean on eastern North America is limited by the fact that the prevailing winds are westerly. Mean annual precipitation amounts are greatest along the coastal zones and decrease as one moves inland from the

Pacific and Atlantic coasts (Figure 4.1). There is also an increase in mean annual precipitation as one moves from the northern arctic regions to the more temperate southern latitudes. Snow is the predominant form of precipitation in the more northerly latitudes and in the higher elevations of the Cascade and Sierra Nevada Mountains of the west coast and in the Rocky Mountains of the west central region of the continent. Rain is the predominant form of precipitation over the rest of the continent.

Barry and Chorley (1987) note that there are at least eight major types or seasonal precipitation regimes in North America (Figure 4.2). The Westerlies–Oceanic and Mediterranean regimes along the west coast receive maximum precipitation in winter, the Intermontane Transitional regime receives maximum precipitation in spring, and the continental interior has a maximum during the warmer spring and early summer months. Further north, in the con-

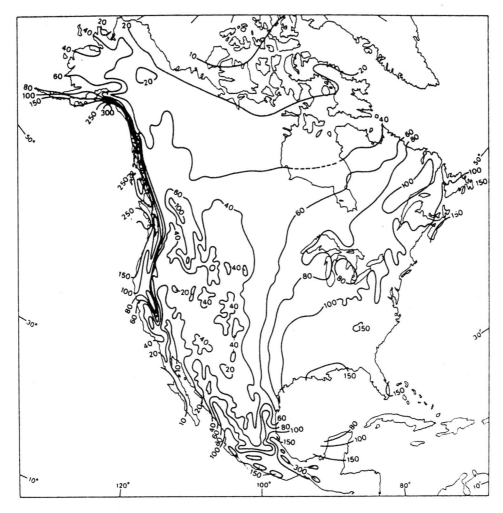

Figure 4.1 Mean annual precipitation (in cm) over North America for the period 1931–60 (data from Barry and Chorley, 1987). Isohyets in the Arctic underestimate the true totals by 30–50% due to problems in recording snowfall accurately.

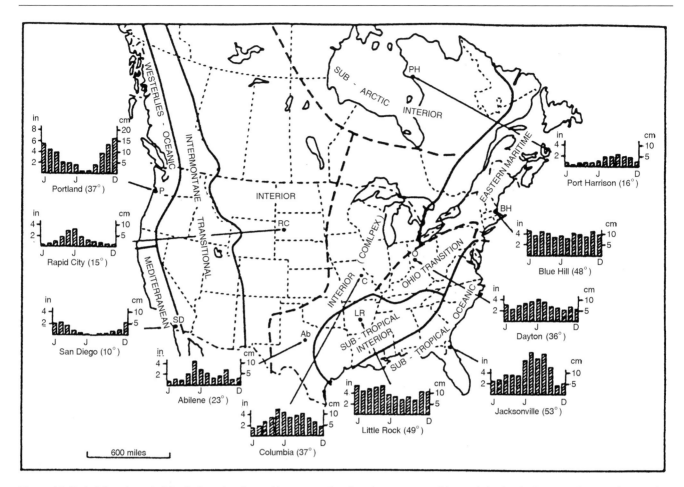

Figure 4.2 Rainfall regimes in North America. Inset: histograms showing the mean monthly precipitation in January, June and December (from Barry and Chorley, 1987).

tinental interior of Canada, the maximum is in late summer and early autumn, and there is a local maximum in early autumn on the eastern shores of Hudson Bay due to the effects of open water. In the Interior (complex) regime there is a double maximum, in May and September. Precipitation is relatively abundant along the east coast in all seasons; the Eastern Maritime regime shows a relatively even distribution throughout the year, while the remaining eastern regimes have summer maxima.

Seasonal patterns of streamflow may be similar or dissimilar to their associated seasonal patterns of precipitation (Figure 4.2). Streamflow varies seasonally as a function of precipitation seasonality, storm type associated with the season, and seasonal evaporation demand. In those regions where precipitation occurs predominantly in the form of rain, streamflow generally reflects a timing response similar to the seasonal distribution of the rainfall. However, in Canada, the northern United States, and in the mountain regions of the continent, precipitation from October to April or May typically falls as snow and is stored in snow-

packs for later release as snowmelt and subsequent stream-flow. Storage periods range from a few days in transitional rain and snow zones, to 4–6 months in the northern latitudes and higher mountain elevations. In the major snow regions of North America, snowmelt begins between April and June depending on the latitude, and may contribute 70–90% or more of the annual streamflow over a few weeks or months.

Winter season rainfall is typically frontal in origin, whereas warmer season rainfall is typically convective in origin. Frontal storms can produce large amounts of rainfall over several days creating floods of several days duration while the summer convective storms may also produce floods of large magnitudes but lasting only a few to several hours. Exceptions to this in the summer are evidenced by such storm periods as the one that generated the flooding on the Upper Mississippi River in 1993 (U.S. Department of Commerce, 1994). Heavy precipitation occurred for extended periods of time with precipitation amounts for June to August being 200–350% of normal for the region.

Tropical storms and hurricanes are also sources of heavy rainfall and major flooding. These storms affect the east and west coasts of Mexico and the east coast of the United States and southern Canada.

Evaporative demand also plays a major role in the determination of hydrological regimes. Evaporation normally reduces the amount of precipitation that is available for a variety of hydrological processes, including groundwater recharge and streamflow. Through its effects on soil moisture, evaporation also exerts a major influence on the type and areal distribution of the ecosystems of North America. The difference between precipitation and evaporation provides a measure of an important climate–hydrology relation. The distribution of the difference between annual precipitation and open-water evaporation for North America is shown in Figure 4.3.

Areas where annual precipitation exceeds evaporation demands typically are more humid and have perennial-type streams as the result of sufficient rates of groundwater recharge to support streamflow. These areas include the eastern and northern regions of the continent plus southern Mexico. Areas where annual evaporation demand exceeds annual precipitation are typically more arid and have more intermittent-type streams due to insufficient rates of groundwater recharge to support streamflow. These areas include the central and southwestern regions of the continent. Changes in this distribution of precipitation–evaporation resulting from the variability of or changes in climate variables such as precipitation, temperature, and solar radiation, could have significant impacts on water-related resources. These include changes in water availability, the distribution of natural ecosystems, and the viability of a range of current agricultural and water resources management practices.

The temporal and spatial variations in the magnitudes of annual and seasonal values of precipitation, temperature, and streamflow in North America have also been shown to be related to large-scale atmospheric circulation patterns. Two of these are the Southern Oscillation and the Pacific–North America (PNA) patterns (Redmond and Koch, 1991). The Southern Oscillation is characterized by a gradient in surface pressure between the eastern and western subtropical Pacific just south of the equator. Periods when the east–west pressure gradient is weak are often associated with a condition of higher than average sea surface temperatures in the equatorial waters off South America, which is termed the El Niño. Together, these two events are termed the El Niño–Southern Oscillation (ENSO). The PNA pattern consists of a deeper than usual area of low pressure over the Aleutians, an intensified ridge of high pressure over western North America, and a negative pressure anomaly at upper levels of the southeastern United States.

Redmond and Koch (1991) determined the nature and magnitude of the relationships between precipitation, temperature and streamflow in the western United States, and ENSO and the PNA index. Statistically significant differences in precipitation occurred during the extremes of the Southern Oscillation index (SOI). During an ENSO event, precipitation was found to be low in the Pacific northwest and high in the desert southwest, and streamflow showed associations with the SOI similar to precipitation. Strong associations for precipitation and temperature with the PNA were also noted in the Pacific northwest.

Kahya and Dracup (1993) found a strong relation between streamflow and ENSO events for four core regions of the United States: the Gulf of Mexico, the northeast, the north-central region, and the Pacific northwest. Dracup and Kahya (1994) found evidence of a streamflow response to La Niña events in these same four regions, but with the sign of the seasonal streamflow anomaly being opposite to that associated with El Niño events. The relationships between streamflow and La Niña/El Niño events were statistically significant based on the hypergeometric distribution and demonstrate the connections between the tropical SOI and streamflow in mid-latitude regions of the United States.

Long-term fluctuations in winter atmospheric circulation over the North Pacific and North America, that are not immediately distinguishable from natural atmospheric variability, were determined to be related to a trend towards warmer winters in central California (Dettinger and Cayan, 1995). This warming has an associated trend of earlier snowmelt runoff in mid-altitude basins where a smaller fraction of runoff occurs in April–June, and a compensating increasing fraction occurs at early times in the water year (October–September).

Relationships between atmospheric circulation and the spatial and temporal distributions of snowpack accumulations in the western United States were demonstrated by McCabe and Legates (1995). Their results showed that winter mean 700 h Pa height anomalies account for a statistically significant portion of both the spatial and temporal variability in snowpack accumulations as measured on or about April 1.

Following up on this work, McCabe and Dettinger (1995) used the Geophysical Fluid Dynamics Laboratory (GFDL) GCM to simulate the 700 h Pa anomalies over North America and the winter precipitation at eight locations in the contiguous United States, and compared them with corresponding correlations in the observed data. The

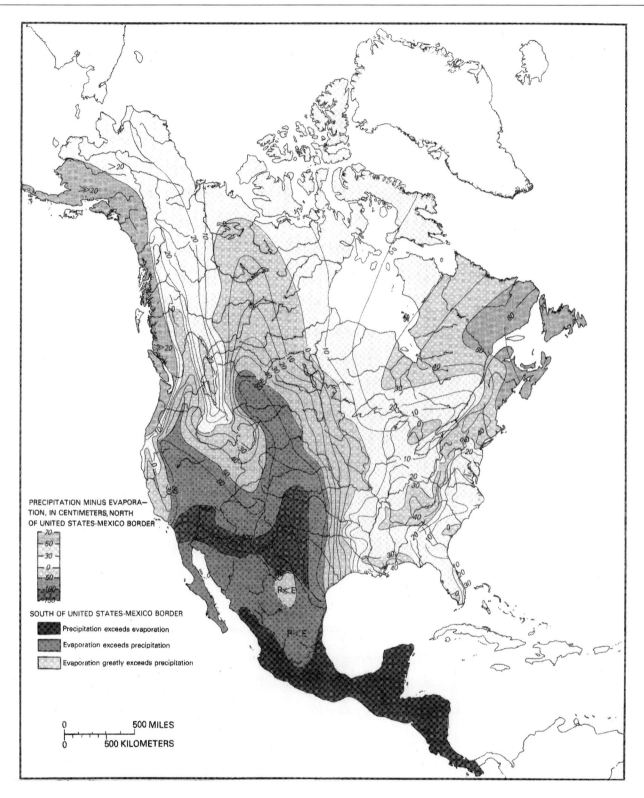

Figure 4.3 Distribution of the difference between precipitation and open-water evaporation in North America (from Winter, 1989).

results indicated that there were some similarities in the simulated and observed relations for most of the sites. However, the simulations of these relations were most similar to observations for locations near oceanic sources of atmospheric moisture. The authors noted that the results suggest that the GFDL GCM may not adequately simulate variations in the advection of atmospheric moisture into the interior of the United States, and/or that moisture is not adequately converted to precipitation in the interior of the continent.

4.3 HYDROLOGICAL CLIMATE CHANGE IMPACT STUDIES IN NORTH AMERICA —

Concerns regarding the effects of climate variability and change on the hydrology of North America have generated a variety of studies over a range of spatial and temporal scales. Some studies have focused on hydrological variability and its response to climate change, while others have focused on selected climate variables, such as precipitation and temperature, and the ability of climate models to adequately simulate the spatial and temporal distribution of these variables. Some investigations have examined the entire continent or large regions of the continent, while others have covered selected regions.

This section provides a brief overview of these studies and their findings. The larger-scale studies are discussed first, followed by the more regional investigations. For the purpose of regional discussions, North America has been divided into eight regions, as defined by the Symposium on Regional Assessments of Freshwater Ecosystems and Climate Change in North America (McKnight and Weiler, 1995; see Figure 4.4). Detailed descriptions of these regions and discussions on the potential impacts of a doubling of CO_2 on freshwater ecosystems of each region is to be published in a special issue of the journal *Hydrologic Processes*.

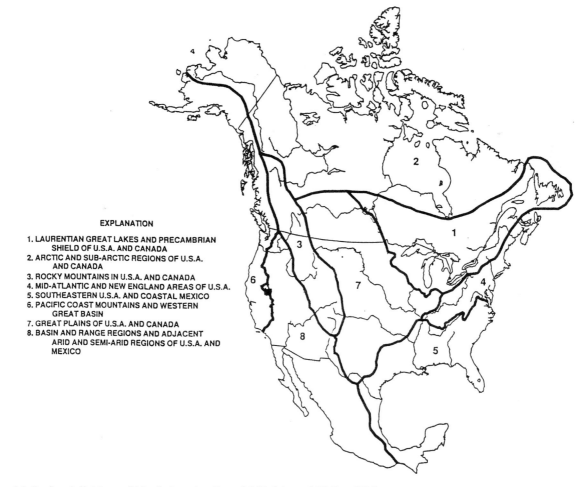

EXPLANATION

1. LAURENTIAN GREAT LAKES AND PRECAMBRIAN SHIELD OF U.S.A. AND CANADA
2. ARCTIC AND SUB-ARCTIC REGIONS OF U.S.A. AND CANADA
3. ROCKY MOUNTAINS IN U.S.A. AND CANADA
4. MID-ATLANTIC AND NEW ENGLAND AREAS OF U.S.A.
5. SOUTHEASTERN U.S.A. AND COASTAL MEXICO
6. PACIFIC COAST MOUNTAINS AND WESTERN GREAT BASIN
7. GREAT PLAINS OF U.S.A. AND CANADA
8. BASIN AND RANGE REGIONS AND ADJACENT ARID AND SEMI-ARID REGIONS OF U.S.A. AND MEXICO

Figure 4.4 Regional divisions of North America (from McKnight and Weiler, 1995).

4.3.1 Continental scale studies

The relations between runoff and climate variability for major river basins of the United States have been examined by a number of researchers. In one of the early studies, mean annual runoff from 22 basins across the United States was related to mean annual total precipitation and a weighted temperature value (Langbein et al., 1949). This relation was used by Stockton and Boggess (1979) to estimate the effects of climate change on 18 water regions of the United States. For a 2°C warming and a 10% reduction in precipitation, they found the largest impacts on basins in the drier western regions and much smaller effects on basins in the more humid eastern regions. Exceptions in the west were the Pacific Northwest and Great Basin regions where it was determined that basins would be less severely affected.

The vulnerability of these same 18 water regions to climate variability and change was also evaluated using five quantitative measures of regional vulnerability (Gleick, 1990). These measures included the ratio of reservoir storage volume to the quantity of supply; the ratio of consumptive use to renewable supply; and the ratio of the flow quantity that is exceeded only 5% of the time to the quantity exceeded 95% of the time. Warning thresholds established by the author were then compared with the measures computed for each region. The Missouri River basin and California exceeded four of the five warning levels, and the Great Basin region exceeded all five warning levels. The findings for the Great Basin were opposite to those of Stockton and Boggess (1979).

Revelle and Waggoner (1983) used Langbein's model to investigate the effects of climate change on runoff and water supplies in the western United States. Their work showed that the effects of a 2°C rise in temperature on decreased runoff was greater than the effects of a 10% decrease in precipitation for areas with a mean annual precipitation of less than 500 mm. Decreases in water supply for the regions studied ranged from about 76% in the Rio Grande region to almost 40% in the Upper Colorado region.

An analysis of measured decadal climate fluctuations and associated runoff changes in 82 basins across the United States over a 50-year period indicated that contrary to the findings of Revelle and Waggoner, temperature fluctuations were not as great a factor in runoff changes (Karl and Riebsame, 1989). In this study, precipitation changes were found to be amplified from one to six times in relative runoff changes, while average temperature changes of 1–2°C often had little effect on annual runoff. The precipitation-runoff change amplifications were considered to be of particular concern because most GCMs suggest that drying will occur over much of the interior of the United States.

Although GCM estimates of average global warming range from 1.8 to 5.2°C, this increase varies in space and time in all models (Mitchell, 1989). The zonally averaged warming is greatest at high latitudes, thus giving Canada potentially more warming than other regions of North America. For the Canadian region, the Goddard Institute for Space Studies (GISS) GCM (Hanson et al., 1984) predicts increases in surface air temperatures of 6–12°C in winter and 2–4°C in summer. The National Center for Atmospheric Research (NCAR) GCM (Washington and Meehl, 1984) predicts increases of 3–7°C in winter and 0–3°C in summer (Mitchell, 1989). Changes in precipitation over Canada are much more variable than changes in temperature and show increases and decreases depending on location and season.

Increases in average winter temperatures will be accompanied by increased winter snowfall in the zone 50–75° N (Manabe and Stouffer, 1980). This may result in a heavier but shorter snow season, increased winter cloudiness, increased annual runoff from low-lying areas, and increased activity of glaciers in mountainous and polar regions (Roots, 1989).

Precipitation seasonality over the United States and the ability of selected GCMs to reproduce this seasonality spatially was examined by Finkelstein and Truppi (1991) using a 90-year record. Variation of seasonality showed that seasonal precipitation patterns are fixed in some areas but change considerably in others. The West coast, the Upper Midwest, and the Southeast Atlantic coast were areas of low variability, while the intermontane west and a large area of the south and east (extending from east Texas to Maine) showed high variability. The patterns of high variability appeared to lie in the transition regions in which orographically influenced weather patterns can be important.

A comparison of control runs for the GISS, GFDL (Manabe and Weatherald, 1987), United Kingdom Meteorological Office (UKMO; Wilson and Mitchell, 1987), and the Oregon State University (OSU; Schlesinger and Zhao, 1987) GCMs with these seasonality data indicated that the models were able to match some important aspects of precipitation seasonality over the United States. These included the West Coast area winter maximum and the area along the Atlantic Coast. However, the large area of convective summer precipitation in the central part of the United States was not well represented. For model runs with a $2 \times CO_2$ scenario, the one common denominator among all models was an increase in the area associated with maximum precipitation in spring.

The projected climate conditions in the United States under conditions of a doubling of CO_2 were determined and compared for the GISS, GFDL, OSU and UKMO GCMs (Feddema and Mather, 1992). In all four models the spatial distributions of temperature changes were quite consistent, but there was some variation in the magnitude of the average deviation from current climate. In all models, each $0.5 \times 0.5°$ grid square was predicted to experience an increase in mean annual temperature. The smallest temperature increases generally occurred in the southeastern part of the country.

The projected monthly precipitation showed much greater variability. All four models showed considerable increases in the nationwide average precipitation, ranging from 34 mm (OSU) to 71 mm (UKMO). Increases in precipitation were generally predicted for the northwestern, north-central, and southeastern parts of the country, while drying was predicted along the Mississippi River valley. Among the model results the greatest discrepancies were found for the southwest and northeast regions.

4.3.2 Regional scale studies

Laurentian Great Lakes and the Precambrian Shield

The effects of global warming on the water resources of the Great Lakes basin were investigated by Cohen (1986). Climate scenarios developed from the GISS and GFDL GCMs were used to generate monthly precipitation and temperature at model nodes within the basin. These values were used as inputs to a water balance model to compute basin runoff to the St Lawrence River downstream from Lake Ontario. Under present wind conditions, the GISS scenario projected a 20.8% decrease in outflow and the GFDL scenario a 18.4% decrease. The results reflect the effects of increased temperatures, which produce relatively high lake evaporation and decreased land-based runoff.

Using just the GISS scenario, Cohen and Allsopp (1988) examined the effects of warming on the province of Ontario with emphasis on the Great Lakes region. Under the GISS scenario, Lake Erie water levels were projected to fall by 44–68 cm; this would have a significant impact on wetlands and associated aquatic habitats.

Global warming would also effect the duration and extent of ice cover on the Great Lakes (Assel, 1991). Three 30-year steady state $2 \times CO_2$ scenarios generated from the GISS, GFDL and OSU GCMs were used as inputs to statistical ice cover models for Lakes Superior and Erie. For all scenarios on both lakes, the duration of ice cover is shorter and the ice cover is less extensive. This could significantly increase winter lake evaporation.

Arctic and subarctic regions

Large areas of the Arctic and subarctic regions are underlain by permafrost. The upper layers of permafrost are sensitive to variations in mean annual temperature. Examinations of ice core samples taken from permafrost in northwestern Canada and Alaska indicate there has been a regional surface warming of 2–4°C over the past century (Lachenbruch and Marshall, 1986). The areas presently underlain by permafrost that are most likely to be affected by global warming are the flat-lying, poorly drained tundra regions (Roots, 1989).

An increase in the thickness of the active layer would increase the area of wetlands. From a hydrological standpoint, increasing the annual thaw depth of permafrost affects the amount of soil moisture storage, the depth to the water table, and associated biogeochemical processes (Kane et al., 1992). The effects of regional warming as predicted by GCMs on permafrost could also be quite significant, with permafrost boundaries moving poleward by about 500 km, thus reducing the area of permafrost to less than 80% of its present extent (Woo et al., 1992). Increased temperatures could also cause these peatlands to dry out, releasing large quantities of methane, which would enhance emissions of greenhouse gases.

The effects of warming on the hydrological regime of an Arctic watershed were investigated by Hinzman and Kane (1992) using a heat conduction model, a hydrological model, and three possible climate warming scenarios. The results indicated that with a 4°C temperature rise, the depth of thaw could double. With a 5°C increase, indications were that an unfrozen zone over, under or within the permafrost, could form and allow subsurface flow to persist throughout the winter. Warming would produce earlier spring melting, greater evaporation, greater soil moisture storage, and a decrease in cumulative runoff. Adding a 15% increase in precipitation to the warming scenario increased soil moisture, cumulative evapotranspiration, and cumulative runoff. A 15% decrease in precipitation had the opposite effect.

With regard to glaciers, warming may be expected to increase melt rates and to reduce glacier size. However, increased warming by a few degrees in the winter months on Wolverine Glacier in southern Alaska produced a period of glacier growth (Mayo and Trabant, 1984). Increased winter warming is associated with moist southerly airflow off the Gulf of Alaska, and the warmer air can carry more moisture into the mountains. Thus the seasonality of warming will be a major factor in determining the effect on glacier mass balances over the rest of Alaska.

Rocky Mountains in the USA and Canada

The sensitivities of streamflow and water supply in the Colorado River Basin to potential changes in climate were investigated by Nash and Gleick (1991, 1993). Hydrological impacts were evaluated by applying the U.S. National Weather Service hydrological model NWSRFS to selected sub-basins of the Colorado using hypothetical temperature and precipitation scenarios, and scenarios developed from the outputs of selected GCMs (GISS, GFDL and UKMO). The effects on water supply were evaluated by applying these hydrological changes to the Colorado River System Simulation (CRSS) model developed by the U.S. Bureau of Reclamation.

The effects of hypothetical changes in temperature and precipitation on hydrology showed that (1) declines in mean annual runoff due to temperature increases alone were 4–12% for an increase of 2°C, and 9–21% for an increase of 4°C; (2) increases in annual precipitation of 10–20% resulted in corresponding changes in mean annual runoff of approximately 10–20%; and (3) a temperature increase of 4°C would require a 15–20% increase in precipitation to maintain runoff at historical levels.

The GFDL-generated climate scenario predicted a large regional temperature increase and no change in precipitation. This resulted in a 10–24% decrease in runoff in the basins studied, the largest decrease in runoff of the three GCM scenarios. The UKMO and GISS-generated scenarios, assuming 20–30% increases in precipitation, respectively, produced increases in runoff of 0–10%.

Temperature increases shifted the seasonal runoff pattern, with an increase in winter runoff and a decrease in spring runoff. This shift resulted primarily from an increase in winter rain and a decrease in winter snowfall. Basins at high elevation appeared to be more sensitive to changes in precipitation and temperature than those at low elevation. In general, runoff in the Upper Colorado basin was found to be more sensitive to a 10% change in precipitation than a 2°C change in temperature. These changes in hydrological responses had significant impacts on water supply parameters such as salinity, reservoir levels, deliveries to users, and hydroelectricity generation.

The sensitivity of runoff in the Animas River basin in southwestern Colorado was examined for a 10% increase in precipitation and potential evapotranspiration, and a 2°C increase in air temperature (Schaake, 1990). As with the previous study, these changes were applied to the U.S. National Weather Service hydrological model (NWSRFS). Average annual runoff increased by about 20% for increased precipitation, and decreased by about 9% for the increased potential evapotranspiration and air temperature. The results showed increased winter runoff, reduced summer runoff, and large decreases in summer soil moisture similar to those of Nash and Gleick (1991, 1993).

The effects of hypothetical climate change on the East River basin in Colorado were evaluated using a range of temperature and precipitation changes (McCabe and Hay, 1994). The results showed that changes in precipitation had a greater effect on runoff than changes in temperature. The effects of a 4°C increase in mean annual temperature could be offset by increases in annual precipitation of 4–5%. The analysis of a gradual increase in temperature of 4°C per century, and a gradual decrease in precipitation of 20% per century, indicated that it would take 80–90 years to have at least a 50% chance of detecting a significant decrease in annual runoff at a 95% confidence level.

Mid-Atlantic and New England areas

Scenarios of climate change developed from the $2 \times CO_2$ runs of the GFDL, GISS and OSU GCMs were used to examine the changes in the Thornthwaite moisture index for the Delaware River basin (McCabe and Wolock, 1992). The GFDL and GISS scenarios produced significant decreases in the index, while the OSU scenario indicated no significant change. The natural variability in climate was also shown to mask the long-term trends in moisture index values, indicating that trends in this index and other hydrological variables could be undetectable for several decades.

Possible changes in the frequency of droughts in the Delaware basin were also examined using a number of hypothetical scenarios of changes in precipitation and temperature (McCabe et al., 1991). Changes in drought frequency were found to be more sensitive to changes in mean precipitation than to changes in mean temperature. Uncertainty in the effects of natural climate variability was also found to have a large effect on the predicted drought frequency.

Southeastern USA and coastal Mexico

The sensitivity of runoff from 52 basins in the southeastern United States was evaluated using a monthly nonlinear water balance model and hypothetical climate change scenarios of

± 10 and ± 20% changes in precipitation and ± 10% changes in potential evapotranspiration (Schaake, 1990). A 10% change in precipitation resulted in changes of 20–45% in mean annual runoff, with the greatest changes occurring in the western part of the region. An increase in precipitation had a slightly larger effect than a decrease. A 10% change in evapotranspiration resulted in changes of 10–34% in mean annual runoff, with the largest effects also occurring in the western part of the region. A decrease in evapotranspiration had a slightly larger effect than an increase. For a given percentage change, changes in runoff were larger for changes in precipitation than for changes in potential evapotranspiration.

Pacific Coast mountains and the Western Great Basin

The Sacramento River basin is an important source of water for California, supplying over 30% of the state's total runoff. Ten scenarios with hypothetical temperature and precipitation changes, and eight scenarios with changes in temperature and precipitation, generated using data from the NCAR, GFDL, and GISS GCMs, were used to evaluate the sensitivity of the Sacramento River basin (Gleick, 1987b). The scenarios were evaluated using a monthly water balance model developed by Gleick (1987a). The results were consistent through all the scenarios, showing reduced summer runoff, increased winter runoff, and summer soil moisture decreases ranging from 8 to 44%. A 20% increase in precipitation resulted in increases of 40–80% in winter runoff for increases of 2 and 4°C. With the GCM scenarios, average winter runoff increases ranged from 16 to 81% and average summer runoff decreases ranged from 40 to 68%.

Similar results were found in an investigation of the sensitivity of three selected sub-basins of the Sacramento River and of one sub-basin of the adjacent San Joaquin River basins (Lettenmaier and Gan, 1990). Climate change scenarios were computed from the $2 \times CO_2$ runs of the GFDL, GISS and OSU GCMs. In these basins, snow is the major source of runoff, and the responses to all scenarios were dominated by temperature-related changes in snowmelt. Increased rainfall in winter resulted in decreased average annual snow accumulation and consequently reduced runoff in late spring, summer and fall.

Great Plains of the USA and Canada

The potential impacts of climate warming on soil moisture, snowmelt, and runoff in the Prairie provinces of Canada were examined by Cohen et al. (1989) and Cohen (1991) in

a study of the water resources of the Saskatchewan River. The headwaters of this basin arise in the Canadian Rocky Mountains in the province of Alberta and the basin extends eastward to near Lake Winnipeg in Manitoba. A total of 15 climate scenarios were evaluated; ten hypothetical scenarios and five scenarios from the GISS, GFDL and OSU GCMs. Runoff was computed using a monthly water balance model driven by the climate scenarios.

The results for the GCM scenarios showed little consensus on the direction and the magnitude of change given the differences in their predictions. The GISS model scenario produced a slight increase in runoff for streams originating in the plains, and a 32% increase in runoff from the mountain streams, resulting from greater snowmelt. Similar results were obtained from the hypothetical scenario of + 2°C with a 20% increase in precipitation.

The GFDL model scenario produced lower runoff from streams on the plains and in the mountains. None of the hypothetical scenarios matched the mountain response, but the + 4°C scenario with a − 20% precipitation provided a similar runoff total for the entire basin. The OSU scenario produced total basinwide runoff that was similar to that measured for the period 1951–80.

Seasonal changes in components of the hydrological cycle were also examined. The GISS scenarios produced wetter winters and drier summers. The GFDL scenario produced relatively wet winters but smaller soil moisture deficits, indicating relatively more humid summers. The OSU scenario produced seasonal changes similar to those of the GISS, but the increased runoff was cancelled out by increases in water consumption in summer.

A study of the sensitivity of semi-permanent prairie wetlands to climate change showed that increased temperatures of 2 and 4°C resulted in significant lowering of water levels and a more frequent drying of the region (Poiani and Johnson, 1993). In most of the simulations, increases in precipitation of 10 and 20% did not compensate for the increase in evapotranspiration related to increased temperature. Groundwater levels in the region were also lower as a result of increased evapotranspiration.

The responses of lakes in Minnesota to changes in climate derived by perturbing historical temperatures with $2 \times CO_2$ GISS GCM outputs for the region were examined by Hondzo and Stefan (1993). The seasonally averaged increase in water temperature in the epilimnion was 3°C, compared with the 4.4°C increase in air temperature related to climate change. Evaporative heat and water losses increased by about 30%, representing increased water losses of about 300 mm/year.

The effects of climate change on the frequency of heavy rainfall in this region were investigated in a study of the state of Illinois using spatial and temporal analogues (Changnon and Huff, 1991). The spatial-temporal analogues provided comparative differences in precipitation and temperature similar to the magnitude of changes obtained from GCM estimates. Warmer and wetter conditions were accompanied by an increase of 10–15% in the frequency of heavy rainstorms with return intervals of 5–50 years. Warm and dry periods had the effect of suppressing the frequency and intensity of severe storms.

Basin and range regions, and adjacent arid and semi-arid regions of the USA and Mexico

The decreases in runoff from semi-arid regions in the United States estimated by Revelle and Waggoner (1983) were questioned by Idso and Brazel (1984), who noted that the previous authors had failed to consider the antitranspirant effect an increase in CO_2 would have on vegetation. Instead of a 40–75% reduction in runoff, their results indicated a possible increase of 40–60%.

Flaschka *et al.* (1987) noted that Idso and Brazel had failed to consider that most of the runoff in the Colorado River basin is derived from snowmelt in the mountains that cover 20% of the basin area. Thus, the runoff generated is relatively independent of transpiration. Using a monthly model to investigate four basins in Nevada and Utah, Flaschka *et al.* (1987) found that the most probable changes in climate – a 2°C increase in average annual temperature and a 10% decrease in precipitation – would reduce runoff by 17–28%.

The question of whether runoff would increase or decrease in this region was further explored using a detailed hydrological model that considers the effects of temperature, precipitation, and CO_2-induced plant function characteristics (Skiles and Hanson, 1994). For small basins in Idaho, Colorado and Arizona, they found that if CO_2 increases with no climate change, a slight increase in runoff may result. If both CO_2 and temperature increase, small increases in runoff may result. However, if precipitation decreases and CO_2 and temperature increases, there would be no short-term increase in runoff.

4.4   NATIONAL AND INTERNATIONAL RESEARCH PROGRAMMES

Although the research efforts cited above have provided some new insights and have improved our understanding of climate–hydrology relations in North America and of the potential sensitivities of regional water resources under a variety of climate change scenarios, they also point out a number of gaps in our knowledge of processes and in the adequacy of the methodologies and models being applied. The wide range of climate change scenarios used in these investigations and the equally wide range of hydrological models and study assumptions applied makes the development of a consistent view of the potential effects of climate variability and change across North America difficult at best.

The ability to estimate the effects of climate variability and change on hydrology is strongly dependent on the development of robust models that can adequately simulate atmospheric and hydrological processes over a broad range of conditions and of spatial and temporal scales. Two basic impediments to the development and testing of such robust models are (1) a limited understanding of some atmospheric and hydrological processes and their interactions over a range of spatial and temporal scales; and (2) the lack of data to improve this understanding and to develop and test the associated models of these processes. To address these problems, a number of national and international programmes have been designed to collect the needed data at a range of scales, and to bring together a critical mass of research scientists, working as a multidisciplinary team, to develop the required understanding and models. Brief reviews of a selected number of these research programmes now being conducted in North America are presented in the following.

4.4.1 The Earth Observing System (EOS)

To address the basic lack of information regarding the earth and climate systems, the U.S. National Aeronautics and Space Administration (NASA) and several other U.S. government and international agencies have developed the 'Mission to Planet Earth' programme, which consists of a series of scientific and flight opportunities known collectively as the Earth Observing System (EOS; NASA, 1991). Current EOS research is focusing on the use of existing satellite data, and preparations for the use of new types of data expected from satellite missions preceding EOS. Detailed requirements for future observations are being determined, and numerical models are being developed that can assimilate or help in the interpretation of current and future data sets.

Research is being coordinated nationally with other U.S. agencies and internationally through the International Geosphere–Biosphere Program (IGBP) and the World Climate Research Program (WCRP). Results from the research activities are being used to guide the development of EOS observatories and the EOS data and information system (EOSDIS). A major goal of EOS is to combine the means for making observations and interpreting data, with a long-term scientific research effort. The result will be an information system that can provide the geophysical, chemical and biological information necessary for intensive studies of planet Earth, and for the development of accurate models of the processes that control the environment.

The EOS observatories will provide global observations from near-polar orbits at specific times of day and night. Some instruments are planned for deployment in low-inclination orbits, providing intensive tropical observations and sampling of the full diurnal cycle. NASA and allied operational meteorological agencies in the United States, the European Space Agency (ESA), Japan and Canada are planning a system of five polar platforms operating simultaneously to support the number of instruments needed to supply the proposed scientific measurements. Two of these five platforms will be supplied by NASA as part of EOS. Each observatory is proposed to be replaced at five-year intervals to provide for 15 years of continuous data.

4.4.2 First International Satellite Land Surface Climatology Project (ISLSCP) Field Experiment (FIFE)

It is generally accepted that climate models are a key component in investigations of the effects of climate variability and change on hydrology, but it has also been recognized that these models may contain poor representations of a number of processes that control the exchanges of energy, water, heat, and carbon between the land surface and the atmosphere. It was this recognition, plus the need to better utilize remote sensing as a tool for the development of data fields for model initialization and validation, that led to the initiation and implementation of the multidisciplinary experiment FIFE. The objectives of FIFE were to understand better the role of biology in controlling land surface–atmosphere interactions, and to investigate the use of remote sensing for estimating or inferring selected land surface parameters (Sellers *et al.*, 1992). The investigation of soil–plant–atmosphere models developed for small-scale applications and of methods to apply these models at scales more appropriate to atmospheric models and remote sensing were major components of the FIFE effort.

Field operations of the experiment were conducted from early 1987 to October 1989 on a 15×15 km test site at and around the Konza Prairie Long-Term Ecological Research (LTER) site near Manhattan, Kansas. Satellite, aircraft and ground-based measurements of meteorological, biophysical and hydrological data were collected over this period and during a number of intensive 12–20 day field campaigns conducted in 1987 and 1989.

The experiment provided details of a number of processes involved in energy, heat and mass exchanges. The identification of linearities and nonlinearities in these processes and their interactions have also provided evidence that some of the small-scale features of surface–atmosphere processes can be extrapolated in an aggregated approach to describe large-scale processes and exchanges using remotely sensed data of relatively coarse resolution (Sellers and Hall, 1992).

The results also confirmed the critical importance of soil moisture in the root zone to the surface energy balance. However, it was also shown that microwave remote sensing could provide only limited information on the moisture content of the upper 0–5 cm surface soil layer. A spatially distributed hydrological simulation model (Famiglietti *et al.*, 1992) developed to describe rainfall–runoff relations in a sub-catchment of the FIFE area and the spatial and temporal distribution of soil moisture over the sub-catchment showed good agreement with flux measurements in the catchment. This suggested to researchers that the hydrological model provides a potential alternative or complement to microwave remote sensing of soil moisture. The combined use of remote sensing and modelling is now being explored to infer regional and continental-scale soil moisture fields. A complete discussion of the major results of FIFE was presented in a special issue of the *Journal of Geophysical Research*, **97**(D17), in 1992.

4.4.3 Boreal Ecosystem–Atmosphere Study (BOREAS)

Following FIFE, it was recognized that this type of study could be usefully applied to other climatic and physiographic regions of the world. The next region selected was the boreal forest region, and planning for BOREAS was begun in 1990. The goal of BOREAS is to improve our understanding of the effects of coniferous forests on global climate (Canadian GEWEX Science Committee, 1992; Sellers and Hall, 1994). The exchanges of heat, energy, water and selected gases such as CO_2 between boreal forests and the atmosphere are being studied using methods of data collection and analysis similar to those used in FIFE.

Research is now being conducted at two sites in Canada located near the northern and southern limits of the boreal forest biome. The northern site is at Nelson House, near Thompson, Manitoba, and the southern site is at Prince Albert National Park, Saskatchewan. Each site has an intensive study area of roughly 20×20 km. This size allows the acquisition of useful airborne flux measurements and satellite observations and a reasonable coverage with surface instruments.

The specific experimental objectives of BOREAS include the development and validation of remote sensing algorithms to transfer understanding of land surface–atmosphere exchange processes from local scales out to regional scales, testing the limits and sensitivity of different surface energy and mass (H_2O and CO_2) balance modelling approaches, and the development and evaluation of remote sensing algorithms that relate the parameter drivers associated with these models to spectral, spatial and temporal patterns of surface radiance fields. The types of models to be applied include (1) biophysical models that describe the absorption and partition of energy by the surface on the spatial scale of the flux measurements: (2) carbon budget models that address gains from photosynthesis and losses due to respiration, fire and runoff; (3) mesoscale meteorological models that address a number of flux and scale issues; and (4) hydrological models, operating over the entire range of spatial scales addressed by BOREAS, to provide independent checks on water budgets over longer time scales and to contribute to the 'scaling up' procedure in concert with the remote sensing and mesoscale modelling efforts.

At each site, intensive field campaigns of 10–20 days are being conducted seasonally to take a wide range of meteorological, ecological, hydrological, and remote sensing measurements. Preliminary results from the May–June 1994 work indicate that very little water vapour was released to the atmosphere during that period. Although the system was wet, much of the soil water system was still frozen in layers 30–100 cm deep (Sellers and Hall, 1994).

4.4.4 GEWEX Continental-scale International Project (GCIP)

Programmes such as FIFE and BOREAS are helping to improve our understanding of processes at the relatively small scales of metres to a few tens of kilometres. However, to move to the scales of application of GCMs, there is a need to expand this understanding to the meso- and continental scales. GCIP, an initiative of the World Climate Research Programme (WCRP), has been designed to study the water and energy budgets of an extensive geographical area of the Earth for which a large volume of data is accessible (WMO, 1992a). The major focus of GCIP is to improve scientific understanding and the ability to model, for climate prediction purposes, the coupling between the atmo-

sphere and the land surface on a continental scale. The GCIP activities are focused on the Mississippi River basin.

The GCIP Science Plan (WMO, 1992a) poses a number of scientific questions that need to be addressed to advance our knowledge of the hydrological and energy cycles involved in the complex land–atmosphere–ocean interactions for a major river basin. To address these questions, four programme objectives have been defined:

- to determine the spatial and temporal variabilities of hydrological and energy budgets over a continental scale;
- to develop and validate macroscale hydrological models, related high-resolution atmospheric models, and coupled hydrological–atmospheric models;
- to develop and validate information retrieval schemes incorporating existing and future satellite observations, coupled with enhanced ground-based observations; and
- to provide a capability to translate the effects of future climate change into impacts on water resources on a regional basis.

The effects of scale on the understanding and modelling of physical and dynamic processes over a land surface is a major research component of this programme. The consideration of nonlinear scale interactions to achieve aggregation of smaller processes to the large scale, and vice versa, is a focus for the development of new methodologies to represent the coupling of processes that are important in the atmosphere to those that are important at the land surface. The techniques developed must be suitable at the resolution of operational prediction of GCMs (of the order of 10–100 km), and hence must be capable of representing in aggregate the effects of high levels of heterogeneity in the underlying ground surface. To address these issues of scale, the GCIP research effort will focus its activities on four scales (IGPO, 1994):

- Continental-scale area (CSA) activities will span the entire domain of the Mississippi River basin. The scale size is that of the Mississippi River basin, 3.2 million km^2.
- Large-scale area (LSA) activities will be undertaken in a phased timetable, emphasizing a particular region with special characteristics. Four LSAs have been identified that in aggregate cover most of the GCIP domain (see Figure 4.5). Each LSA will be investigated for a period of about two years.
- Intermediate-scale area (ISA) activities will be phased in step with those for the LSAs and will serve as the basis for the regionalization of the parameters and coeffi-

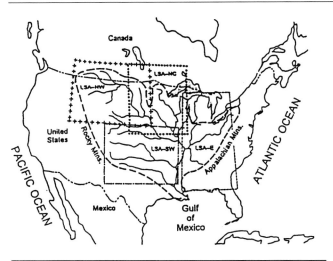

	Year					
	1994	1995	1996	1997	1998	1999
	GIST	ENHANCED OBSERVING PERIOD				
Continental-scale area	**	***	***	***	***	***
Large-scale area						
LSA–SW	***	***	***	**	**	**
LSA–NC	*	*	***	***	*	*
LSA–E	*	*	*	***	***	*
LSA–NW	*	*	*	*	***	***

Emphasis codes: *** high; ** moderate; * low.

Figure 4.5 Boundaries of large-scale areas (LSAs) and temporal emphasis for each LSA from 1994 through 1999 (from Leese, 1995).

cients of land surface hydrology models. The ISA scale is the scale at which hydrological components will be parameterized in atmospheric models. The scale size is of the order of 103–104 km². Activities at the ISA scale will include the analysis of existing basin-scale hydrological models and the analysis of scaling relationships between the LSA and ISA scales.

• Small-scale area (SSA) activities will typically be undertaken in association with efforts requiring intense observing periods over a concentrated region in order to study a focused set of issues. The scale size is less than 102 km².

A number of 'principal research areas' have been defined to ensure that specific scientific studies or development efforts are carried out in a timely manner in order to achieve the GCIP's scientific objectives. These principal research areas and their goals are as follows:

1. *Diagnostic studies:* to achieve a better description and understanding of the annual cycle of hydrological processes over the Mississippi River basin.
2. *Coupled modelling:* to improve the representation of land surface components in GCMs.

3. *Data assimilation:* to improve techniques for using models and data together, in order to take advantage of the information contained in each, and to add value to observed data sets.
4. *Project for the Intercomparison of Land Surface Parameterization Schemes (PILPS):* to improve understanding of the capabilities of land surface schemes in atmospheric models.
5. *Water resources research:* to apply coupled atmospheric and hydrological model outputs as inputs for operational hydrological and water resources management models over a range of time scales using coupled model outputs.
6. *Precipitation:* to improve understanding and estimation of the spatial and temporal structure of precipitation over the Mississippi River basin, including improvements in atmospheric model representations of precipitation to support better coupled modelling.
7. *Streamflow and runoff research:* to describe better the spatial and temporal distribution of runoff over the GCIP study area and to develop mechanisms for incorporating streamflow measurements in the validation and updating of coupled land–atmosphere models.
8. *Soil moisture:* to improve the estimation and understanding of the spatial and temporal structure of soil moisture, and the relationship between model estimates and observations of soil moisture.
9. *Land surface characteristics:* to improve quantitative understanding of the assumed or potential relationships between model parameterizations of land processes and land surface characteristics and to facilitate the availability, testing and evaluation of data and information on land surface characteristics for model development and validation.
10. *Clouds and radiation research:* to improve the descriptions and understanding of the physical and radiative properties of clouds and their representation in atmospheric models.

Studies are currently ongoing in LSA southwest (SW), with emphasis on the Arkansas and Red River basins of this region. New efforts are planned to begin in the LSA north-central (NC) in late 1996 and in the LSA east (E) in late 1997.

4.4.5 Mackenzie GEWEX Study (MAGS)

The Canadian GEWEX programme is designed to contribute to the international GEWEX Programme and to improve understanding and predictions of likely changes in Canada's water resources due to climate change (Canadian

GEWEX Science Committee, 1992; Lawford, 1992). A major goal of the programme is to provide information to be used in improving the models, especially for the colder, high-latitude regions, and to foster research and related activities aimed at assessing the possible effects of climate variability and change on Canada's water resources.

The Mackenzie GEWEX Study (MAGS) uses the Mackenzie River basin, a cold region basin with a northward flowing river system, as a principal component of the Canadian GEWEX programme. The Mackenzie basin will be the focal point for process studies and the development of hydrological models. The basin meets most of the GEWEX–GCIP basin selection criteria, although the existing database is poor. Within the basin there are many important cold region phenomena such as snow–ice processes, permafrost, ice jams, and cloud–radiation interactions that will be essential components of any global climate system model. Taken together, the Mackenzie and Mississippi watersheds provide a true continental area in which to test and validate a GCIP macroscale hydrological model. In addition, there is only limited regulation of the Mackenzie River, and it plays a significant role in determining the degree of ice cover and circulation of the Arctic Ocean.

A central goal of MAGS is to develop the ability to model the water and energy balances of the Canadian Arctic at a spatial scale of 100 km and a temporal scale of one month. Other objectives are (1) to improve land surface process modelling within climate system models by developing a hydrological model that can be embedded in atmospheric climate models, and (2) to estimate present and possible future surface water balances of the Canadian Arctic at GEWEX scales, using existing databases and GCM simulations. Studies of hydrological processes will focus on snowpack development and removal, interception by vegetation, snowfall measurements, evapotranspiration, permafrost hydrology, and glaciated basins.

Selected mesoscale basin models will be applied to a small number of tributaries of the Mackenzie for which hydrometric records are available. The results of the process studies will be incorporated in these model formulations. Critical elements will include parameterization of permafrost effects, snow sublimation, glacier processes and mass balance evapotranspiration. The models will be tested by driving them with the outputs of numerical weather prediction models or GCMs in order to test their responses to time-varying meteorological inputs. The results will be compared with observed streamflow or other observations of the basin. A major focus of the modelling research will be the issues of scaling, transferability and applicability.

Preliminary assessments of the water budgets of the Mackenzie Basin have been made, and work is ongoing to complete the identification of key processes, background climatology, and improved representations of physical processes in selected models (Krauss, 1996).

4.4.6 Arctic Climate System Study (ACSYS)

Processes in the Arctic region have considerable effects on the global climate. The Arctic Ocean plays an significant role through its influence on the thermohaline circulation of the world's oceans by exporting sea ice and fresh water to lower latitudes, and through the seasonal changes in albedo associated with changes in snow and sea ice cover. ACSYS was initiated to improve understanding of climate processes within the Arctic basin, as well as linkages with Arctic region hydrology (WMO, 1992b).

A major goal of ACSYS is to provide a valid scientific basis for the representation of the Arctic region in coupled global atmosphere–ocean models. A major component in accomplishing this goal is the development, testing and refinement of macro-scale hydrological models that can be used to provide the continental link in the climate feedback loop among global atmospheric circulation, precipitation, evaporation, runoff, and the buoyancy forcing of the Arctic Ocean. The Arctic Ocean contains only 1.5% of the world's ocean water, but receives about 10% of the total global freshwater runoff. The Mackenzie River and a large number of northern coastal basins are the major contributors of freshwater to the Arctic Ocean from North America, and thus the results of the MAGS effort will make a major contribution to ACSYS. ACSYS is currently in the planning phase; data collection and modelling are proposed to begin in the near future.

4.4.7 Semi-Arid Land Surface–Atmosphere Mountain Experiment (SALSA-MEX)

SALSA-MEX was formulated to measure and predict land–atmosphere interactions in a semi-arid mountainous region and to address the effects of topography on hydrological and meteorological processes over many scales (Goodrich, 1994). In contrast with the short time duration of some previous biome-type studies such as FIFE and BOREAS, SALSA-MEX has been designed to be conducted for a longer period of 3–10 years.

The site selected for this study is the San Pedro basin, which straddles the US–Mexico border in southeastern Arizona and northern Sonora, Mexico. Research will be conducted collaboratively by US and Mexican investigators. The basin area is about 12 000 km^2, with headwaters near Cananea, Sonora in Mexico, and outlet near Winkelman, Arizona in the United States. Major biome types represented are the Chihuahuan desert, semi-arid grasslands, oak savannah chaparral, piñon–juniper, and coniferous forest.

The initial experimental research objectives proposed for SALSA-MEX are:

- To quantify hydrological fluxes and to identify the dominant hydrological processes as functions of temporal and spatial scales in regions with high topographic relief, with particular emphasis on: (a) surface–groundwater interactions in the San Pedro riparian area, and (b) the role of near-surface soil moisture in infiltration and runoff generation.

- To evaluate evaporation and transpiration and their relationships to the water balance in topographically rough terrain over a range of canopy-soil-understorey conditions represented by the biome types in the San Pedro basin.

- To assess the utility of remotely sensed data for regional land surface characterization and for incorporation into hydrological and energy balance models utilized for the first two objectives.

- To test the ability of mesoscale meteorological models to simulate realistically a broad range of land–atmosphere interactions in heterogeneous domains with significant topographic relief.

- To determine the spatial and temporal patterns of carbon uptake and release by the vegetation–soil continuum. In addition to seasonal trends, the effects of water availability, particularly from precipitation, on these fluxes will be evaluated.

- As part of the above objectives, state variables and fluxes over a range of scales will be observed and modelled, in order to formulate scaling relationships for aggregation and disaggregation.

The SALSA-MEX programme was started in 1995 and investigations are currently ongoing.

4.5 SUMMARY AND CONCLUSIONS

The effects of climate variability and change on hydrological regimes in North America have been investigated by a large number of investigators using a wide variety of methodologies and a broad range of assumptions regarding the magnitude and direction of change. The results of some regional studies show a measure of consistency. One example is the shift in the timing of runoff in snowmelt-dominated regions as a result of increased air temperatures, which appear to cause an increase in the proportion of rain to snow during winter months. However, the differences within and among studies in other regions are considerable and the results are much more inconclusive. In addition, comparisons among regions are not always possible, given the lack of consistency in the methodologies and the assumptions applied.

Current studies have typically used a hydrological model driven by one or more climate change scenarios to assess the effects of climate change on the hydrology of a basin or region. Problems in this approach are related to uncertainties in both the climate scenarios used, which are typically generated using results from one or more GCMs, and in the hydrological models used. To reduce the uncertainty in the results will require the development of improved atmospheric and hydrological models, and of consistent and robust methodologies that facilitate inter- and intra-regional comparisons of results.

Efforts are now being made to improve process understanding and to develop improved process models in a variety of national and international studies being conducted in North America. Many of these studies are quite comprehensive in scope, but most are of limited spatial or temporal extent. The results are providing valuable new insights into processes, but only during a single phase of a climate period that may or may not contain any degree of climate variability associated with the region. Longer-term programmes such as EOS could provide the opportunity to observe a wider range of climate variability, but to maximize the benefits of such programmes, they must be coupled with land surface-based studies of comparable duration.

Any increase in the number of climate–hydrology related studies or their duration, however, will be limited to a large degree by current funding levels. Managers of resources related to or affected by the hydrological cycle will therefore have to continue to rely on scientists' 'best estimates' of future climate change for any long-term planning, and will have to await the anticipated improvements in models and methodologies to obtain better estimates.

5 Assessment of the impacts of climate variability and change on the hydrology of Europe

N. W. ARNELL

5.1 INTRODUCTION

The droughts of 1988–92 and the floods of 1994 and 1995, both of which affected many parts of Europe, served as a reminder that climatic and hydrological variability still have significant economic and social impacts, despite decades of investment in a wide variety of water and river management schemes. Inevitably, the question of whether these 'extreme' events were signs of global warming has frequently been raised. It is, of course, too early to draw any definitive conclusions, but the recent floods and droughts have shown the sensitivity of water resources in Europe to change.

There have been several studies into the potential effects of climate change on hydrological characteristics in Europe, using unfortunately a wide range of scenarios and different methods of analysis. This chapter reviews many of these studies and summarizes some of the experiments that have been undertaken in Europe to explore the processes relating climatic variability to hydrological behaviour. First, however, it is necessary to summarize European hydrological regimes.

5.2 HYDROLOGICAL REGIMES IN EUROPE AND THEIR VARIABILITY OVER TIME

In the most general terms, hydrological regimes in Europe can be divided into two types: regimes dominated by rainfall, and regimes dominated by snowmelt. Rainfall-dominated regimes, with maxima in the winter and minima in late summer, occur in the west and south, whereas snow-dominated regimes, with maxima in spring and minima in summer or winter, are found in the north and east. In practice, of course, the picture is considerably more complicated than this (Krasovskaia *et al.*, 1994).

There are significant differences between the rainfall-dominated regimes of western Europe, which are controlled by the passage of Atlantic depressions, and those of southern and Mediterranean Europe. These latter regimes are characterized by a concentration of rainfall in winter, and the occurrence of short-duration, high-intensity events; river flows are therefore much more variable than those further north, both throughout the year and between years. There are also significant variations between different snow-dominated regimes, depending essentially on the relative contribution to total runoff of the spring snowmelt peaks and summer and autumn rainstorm peaks.

Topography adds a complicating pattern to the map of regime types across Europe. There is a great complexity in the regime types in both Scandinavia and the Alps, where rainfall, snowmelt and glacier melt may all contribute to the runoff hydrograph. Finally, many of the large rivers of Europe cross several different climatic zones. Perhaps the best example is the Rhine, which rises in the Alps and hence receives a major contribution in spring from snow and glacier melt, but which also receives large quantities of water from maritime, rainfall-dominated basins such as the Mosel. The Danube also rises in the Alps and receives contributions from tributaries with peaks at different times.

Climatic characteristics in Europe range from semi-arid in parts of southeastern Europe and the Mediterranean fringe, through humid-temperate to humid-cool. The proportion of annual precipitation going to runoff varies from under 10% to over 80%.

Large parts of Europe are affected by the passage of either Atlantic depressions or seasonal anticyclones, so there are very strong spatial patterns in variability over time. Large areas tend to experience similar anomalies at the same time. Figure 5.1, for example, shows river runoff during two winters across northern and western Europe (Arnell, 1994). There is also an apparent tendency for years with above (or below) average conditions to cluster (Arnell *et al.*, 1990; Arnell, 1994). The temporal pattern in Europe is not so obviously clustered or periodic as in some other parts of the world, but there are patterns which merit further investigation. Hisdal *et al.* (1995) examined time series in different

DJF 1971/72 river flows

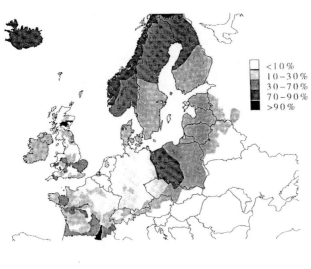

DJF 1979/80 river flows

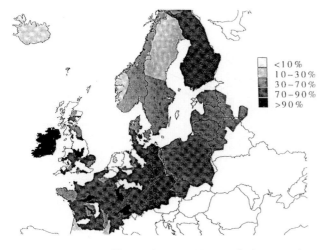

Figure 5.1 River runoff in northwestern Europe during two winters (December–February): 1971/72 and 1979/80 (Arnell, 1994).

parts of the Nordic region, and found few general or consistent trends. There was a tendency for increased runoff in much of the region since the 1980s, but whether or not trends were identified depended to a large extent on the period of record used. Some trends were the result of jumps in the mean, while others reflected a gradual change. The temporal variations that were identified were associated with variations in precipitation, temperature and atmospheric circulation patterns. For the upper Danube, Gilyen-Hofyer (1994) found little change in annual runoff over the period 1883–1985, but a significant decrease in autumn runoff and an increase in peak discharges during summer.

Finally, it is important to recognize that many river flow regimes in Europe are very heavily influenced by human activities. These influences may be deliberate (resulting from reservoir impoundment, river regulation, and abstraction, for example) or inadvertent (due to urbanization or deforestation, for example). The effects of any climatic variability and change must therefore be superimposed in many basins on unnatural hydrological regimes. The water management structures in a basin may mitigate the effects of change, or may exacerbate them.

5.3 HYDROLOGICAL CLIMATE CHANGE IMPACT STUDIES IN EUROPE

Although there are considerable variations among climate model simulations of climate change in Europe (IPCC, 1990, 1992), several reasonably consistent patterns can be seen. Most models simulate an increase in precipitation during winter across most of Europe, and most simulate an increase in precipitation during summer in the north, and a decrease in the south: however, the dividing latitude between summer increase and decrease varies from central France in some models to Scotland in others. The Hadley Centre's coupled ocean–atmosphere model, for example, has been used to simulate the transient effects of a compounded 1% per year increase in CO_2 concentrations (Murphy and Mitchell, 1995). By the time of CO_2 doubling (years 66 to 75 of the experiment, or around 2050 according to the IPCC IS92a emissions scenario), annual temperatures are simulated to have increased by over 4°C in European Russia and the Ukraine, and by close to 2°C in the north and west of Europe. Winter precipitation increases in all areas except Spain, Italy and southeastern Europe, and summer precipitation is reduced south of Scotland and southern Norway. Models with higher regional resolution which are nested within a global model would give greater spatial detail (Jones et al., 1995).

Many of the studies of the effects of climate change on European river flow regimes have been published in the refereed literature, but at present some are available only in the 'grey' literature as reports. This section summarizes the results of some of these studies, and reviews the methods used and the broad conclusions. For the sake of convenience, Europe is rather arbitrarily divided into five regions; western Europe (predominantly rain-fed regimes), northern Europe (snow-dominated regimes), eastern Europe (also snow-dominated), southern Europe (with Mediterranean-type regimes), and Alpine Europe (with the complicating effects of altitude, topography and glaciers).

5.3.1 Western Europe

Climate change impact studies have been completed and published in Belgium, the UK, lowland Switzerland and the Rhine basin.

Belgium

In a series of papers, Bultot and co-workers (Bultot *et al.*, 1988; Gellens, 1991) investigated the sensitivity of components of the water balance and river flows in three small Belgian catchments to climate change. Their studies used a daily conceptual hydrological model, calibrated in each catchment, together with one climate change scenario derived from the outputs of early general circulation models. The adopted climate change scenarios assumed an increase in rainfall during winter (between 8 and 16%) and a slight decrease in summer (between 2 and 3%). Annual potential evaporation was calculated to increase by 9%, following the assumed changes in temperature, humidity, radiation and windspeed. Table 5.1 summarizes some of the key results. The studies found increases in annual discharges, caused largely by increased winter flows (Figure 5.2). Summer flows were simulated to decrease in all but the Dyle catchment, which has a major groundwater component. It was also projected that additional winter recharge would sustain flows during the drier summers. As a result of this increased seasonality in flows, flood frequency increased and low flows became more common.

The United Kingdom

There have been at least three investigations into the effects of global warming on river flows in the UK (and more on water resource impacts). Palutikof (1987) adopted a temporal analogue approach, comparing river flows in ten catchments in warm and cool 20-year periods. The warm and cool periods chosen, based on northern hemisphere average temperatures, were 1934–53 and 1901–20, respectively. For most of the year, rainfall was lower during the warm period than during the cool period; the temperature difference was only 0.3°C, which is small compared with the changes that might occur under global warming. Not surprisingly, given the scenarios used, Palutikof (1987) found a decrease in runoff in most catchments, and a slight increase in two of the catchments in the north and west during the warmer period.

In two studies conducted at the Institute of Hydrology, Arnell and co-workers (Arnell *et al.*, 1990; Arnell, 1992a,b; Arnell and Reynard, 1993) examined changes in river flows

in a number of study catchments, using conceptual hydrological models and both arbitrary climate change scenarios, and scenarios derived from climate model outputs. The first study (Arnell *et al.*, 1990; Arnell, 1992a,b) simulated monthly river flows in 15 catchments in England and Wales using a three-parameter monthly water balance model (from Alley, 1984). Model parameters were determined by calibration, and generally the model fits were characterized as 'adequate' rather than good. The study used a set of arbitrary climate change scenarios, based loosely on the visual inspection of the output from several climate models. The scenarios for change in rainfall assumed an increase throughout the year or an increase concentrated in winter; three potential evaporation scenarios were used, with annual increases of approximately 7 and 15% (two of the scenarios assumed different distributions of increase throughout the year). The study found that the greatest sensitivity to change in rainfall or potential evaporation was in the catchments with the lowest runoff coefficient, in the south and east of England.

More significant than changes in annual runoff, however, were changes in the seasonal distribution of runoff. Three type cases were identified. In the first, found in upland catchments, there are large percentage changes in both winter and summer flows. In such catchments there is currently a fine balance between rainfall and potential evaporation during summer. A change in rainfall and potential evaporation would send the catchment into either a water balance deficit, resulting in a major reduction in river flows, or would make available excess water, leading to an increase in summer flows. In the second type case, found in lowland catchments, there is already a large difference in summer between potential evaporation and rainfall, and a change in the summer water balance has little effect on summer flows. Instead, the greatest percentage change in runoff is in autumn, at the end of the summer deficit season. The third type case occurs in catchments dominated by discharge from groundwater aquifers. Here, changes in winter recharge control the pattern of seasonal flows; flows may be higher even during warmer, drier summers if winter recharge is increased. The study also showed that empirical relationships between climatic variables and hydrological characteristics (based, for example, on regression analysis) could give very different results from those of the water balance model simulations, for the same change scenarios. Sensitivities to change in climate were shown to be highly dependent on the form and the parameter values of the empirical relationship, and such relationships should therefore not be used in climate change impact assessments.

Table 5.1. *Summary of simulated changes in hydrological characteristics: Belgium and Switzerland (Bultot et al., 1988; 1992).*

	Belgium			Switzerland
	Zwalm	Dyle	Semois	Murg
Change in annual precipitation (%)	7.3	6.7	4.7	4.4
Change in annual potential evaporation (%)	9	9	9	10
Change in annual actual evaporation (%)	7	7	8	9
Change in annual runoff (%)	10	7	3	0.3
Change in winter (December–January) runoff (%)	14	9	9	11
Change in summer (June–August) runoff (%)	− 5	4	−19	−14
Change in number of days with soil moisture content more than 60% below saturation	24	11	9	9
Change in number of days with snow lying	− 10	− 14	− 29	− 58

The second study at the Institute of Hydrology (Arnell and Reynard, 1993) developed the first in a number of ways: it used a wider range of catchments, it employed a daily water balance model, it used scenarios based more closely on climate model output, and it considered a wider range of hydrological measures. The five parameters of the daily conceptual water balance model were calibrated in each of 21 study catchments, and in a northern subset of these catchments a simple snowmelt model was also applied. Climate change scenarios were taken from the UK Climate Change Impacts Review Group (CCIRG, 1991) and from the output of the Hadley Centre high-resolution equilibrium climate change simulation experiment (Viner and Hulme, 1993). The CCIRG 'best' rainfall scenario assumed an increase in winter rainfall but no change during summer: the wettest assumed an increase throughout the year, while the driest assumed no change in winter but a decrease in summer. The scenario based on the Hadley Centre experiment varied across Britain, but generally assumed an increase in winter rainfall and a reduction during summer. Several potential

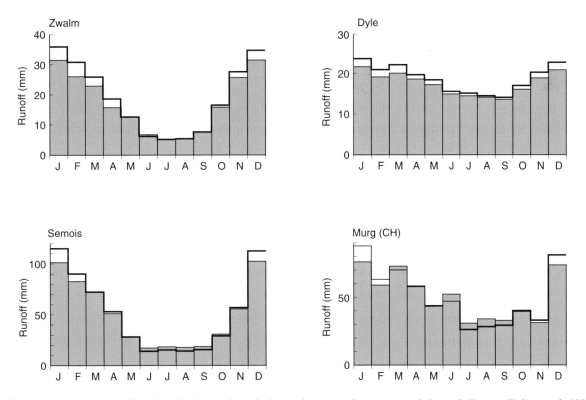

Figure 5.2 Average monthly runoff in three Belgian and one Swiss catchment, under current and changed climates (Bultot *et al.*, 1988, 1992).

evaporation scenarios were produced, making different assumptions about changes in temperature, radiation, humidity, windspeed and plant physiological properties (the assumed changes in temperature and humidity were found to be most important). Annual increases in potential evaporation ranged from 9 to 30%, for an annual temperature increase of around 2.1°C.

Figure 5.3 shows the percentage changes in annual runoff across the whole of Britain (based on gridded climatic input data), and Figure 5.4 shows monthly runoff for six example catchments, under the CCIRG-based scenarios. There is considerable variability in change between the scenarios, but the general conclusions of the first study are supported. Changes in seasonal flow are greater, in percentage terms, than changes in annual discharge, with reductions in summer flow being compensated by increases in winter flow. The percentage changes in shorter-duration low flow measures, such as the flow exceeded 95% of the time, were even larger than changes in seasonal and monthly flows. Snowfall and snowmelt make small contributions to river flows in upland and northern Britain, and under the scenarios considered would be almost entirely eliminated. This occurs largely because of the assumed temperature rise, which would mean that little precipitation would fall as snow.

Finally, the study also considered transient climate change scenarios, assuming a gradual change in climate over the period 1990–2050. Figure 5.5 shows the simulated time series of winter and summer runoff, under one of the CCIRG scenarios, for one study catchment. It is apparent that any trend is small compared with the year-to-year variability; it will therefore be difficult to detect. At the decadal scale, however, there is a clearer climate change trend.

One of the very few studies of potential changes in groundwater recharge has been conducted in the UK. Cooper *et al.* (1995) used a water balance model to estimate percentage changes in recharge, and an idealized aquifer model to translate these into changes in groundwater levels and discharge to streams. It was found that under some of the scenarios considered (which were the same as those used by Arnell and Reynard, 1993) groundwater recharge would increase, but under others it would decrease. The change depended on the extent to which increased winter precipitation was offset by a shorter recharge season caused by increased spring and autumn evaporation.

Lowland Switzerland

Bultot *et al.* (1992) investigated the effects of climate change in the Murg catchment in northern lowland Switzerland,

using the same daily conceptual model and change scenarios as those applied to catchments in Belgium. Some of the key results are shown in Table 5.1 and Figure 5.2; from Table 5.1 it can be seen that the sensitivity of the Murg catchment is similar to that of the Semois in the Belgian Ardennes. Winter flow is increased, summer discharge is reduced, and high and low flow extremes are exacerbated.

The Rhine basin

The Rhine basin, with an area of $250\,000\,\text{km}^2$, is one of the most important transport arteries in Europe, and is a major source of water for public supply, power generation and effluent dilution. Above Basel in Switzerland the river is dominated by snow and glacier melt, and peak flow occurs during summer. Below Basel, flow regimes in the basin are controlled by rainfall and the peak flow period on the Rhine itself occurs earlier with increasing distance downstream.

The potential effects of climate change on river flows in the Rhine catchment have been investigated using the **RHINEFLOW** model, a monthly water balance model that operates across the entire basin on a grid of $3 \times 3\,\text{km}$ (Kwadijk, 1993; Kwadijk and Rotmans, 1995). The model is used to simulate river flows on the major tributaries of the Rhine and at various points along the Rhine itself. Figure 5.6 shows the changes in monthly runoff at three locations, under two of the IPCC 1990 emissions scenarios: the BaU ('business-as-usual') scenario assumes that there will be no explicit response to global warming, while the AP ('accelerated policies') scenario assumes that strict emissions control policies will be introduced.

The spatial distribution of change for each emissions scenario was determined from several climate model simulations. The 'best-guess' results use an average of the different models, while the confidence limits were determined from the spread of output from the different models. In the alpine part of the Rhine basin (upstream of Basel) winter discharges increase and spring flows decline between May and October, due largely to a reduction in snowfall, an increase in glacier melt, and an increase in winter precipitation. In the central part of the basin, flows are simulated to increase between February and July, and to decrease at other times. This pattern reflects not just the change in precipitation through the year, but also the increasing summer evaporative demands. Flows at the Dutch–German border change in response to the altered pattern of runoff in both the snow-dominated and rain-dominated parts of the basin, and the importance of snowmelt on the regime is reduced. Under the 'best-guess' BaU emissions scenario, at the

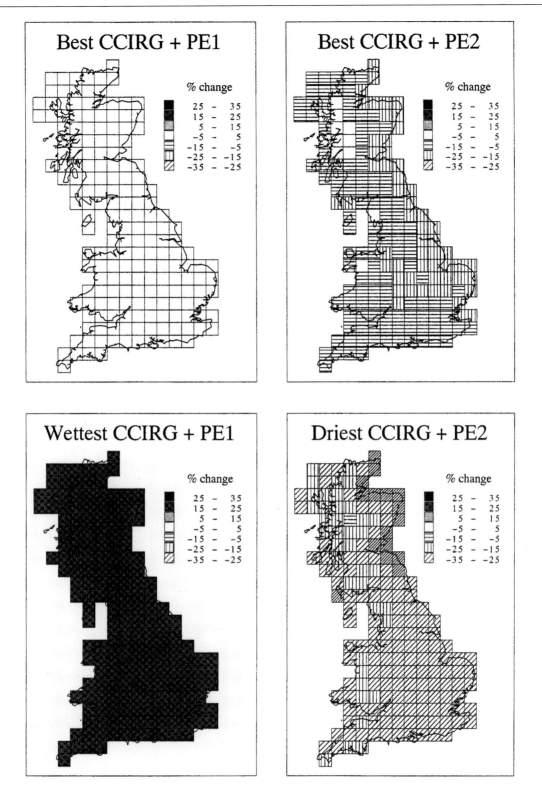

Figure 5.3 Percentage change in annual runoff across the UK by 2050. CCIRG scenarios (Arnell and Reynard, 1993).

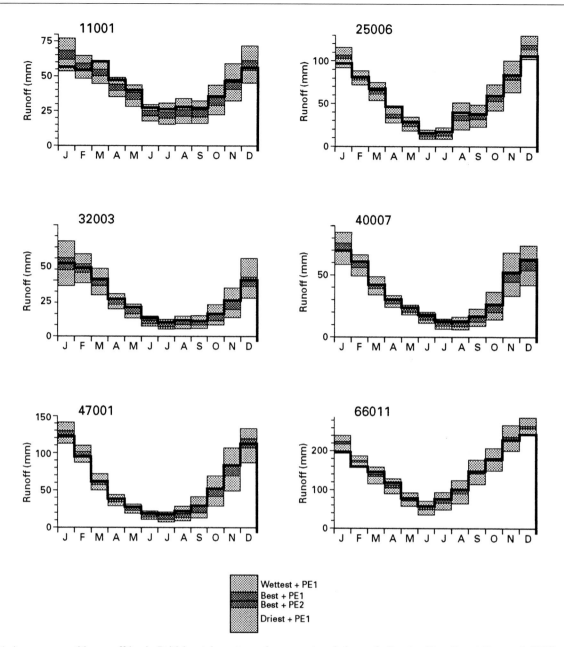

Figure 5.4 Average monthly runoff in six British catchments, under current and changed climates (Arnell and Reynard, 1993).

Dutch–German border monthly flows below 700 m³/s (necessary to control saline intrusion) would increase in frequency from twice in 25 years to eight times in 25 years.

Kwadijk and Middelkoop (1994) developed empirical relationships between monthly and peak discharges to estimate possible changes in flood frequencies on the Rhine. Figure 5.7(a) shows the probabilities of annual peak discharges under a range of precipitation change scenarios, and two scenarios derived from climate model output. One represents the climate by 2100 under the IPCC BaU emissions policy,

and the other assumes an 'accelerated policy' (AP) response. Clearly, at Lobith on the Dutch–German border, the flood frequency curve is very sensitive to climatic variability. The median annual flood (probability of exceedance of 0.5) would increase by around 10% under the BaU scenario. Figure 5.7(b) shows the average number of days per year on which flows at Lobith exceed a threshold of 5500 m³/s; beyond this threshold most of the floodplain of the Rhine is inundated. The frequency of flooding is very sensitive to changes in precipitation.

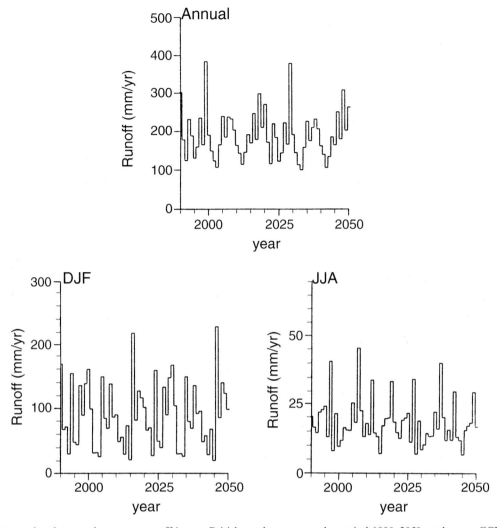

Figure 5.5 Changes in winter and summer runoff in one British catchment over the period 1990–2050, under one CCIRG scenario (Arnell and Reynard, 1993).

5.3.2 Northern Europe

River flow regimes across most of northern Europe are dominated by the effects of snowfall and snowmelt, with the exceptions of the western coast of Norway where rainfall is most important, and southern Sweden where snowfall is moderate.

Sælthun *et al.* (1990) conducted a review of the potential impacts of global warming on Norwegian water resources, which involved the simulation of river flows under altered conditions in a number of representative catchments. The 'probable' scenario assumed an increase in precipitation, particularly in spring and summer, and temperature increases of 1.5–3.5°C. Annual runoff in upland and wet areas was simulated to increase, while runoff would decrease in lowland areas because the increase in evaporation would compensate for additional rainfall. Winter runoff would increase and the spring floods would be reduced in many basins, since less

precipitation is stored as snow. Figure 5.8 shows monthly runoff under current and changed conditions for three elevation bands within the Vosso catchment in western Norway.

Berndtsson *et al.* (1989) used a monthly water balance model to investigate the sensitivity of flows in a small catchment in southern Sweden to arbitrary changes in temperature and precipitation. A temperature rise of 2°C would lead to an increase in annual runoff of 50% if associated with a 20% increase in precipitation, but a decrease in runoff of 65% if precipitation were to decline by 20%.

Similar general patterns were found in 12 catchments in Finland by Vehviläinen and Lohvansuu (1991). Under scenarios derived from the GISS GCM, annual runoff was simulated to increase by 20–50%. Spring discharge peaks, however, would decline substantially, and minimum flows (which currently occur during the winter) would increase considerably. Krasovskaia and Gottschalk (1992) compared

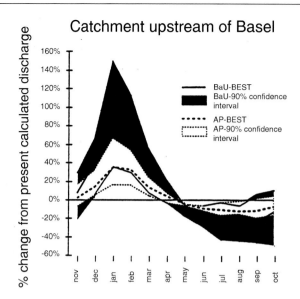

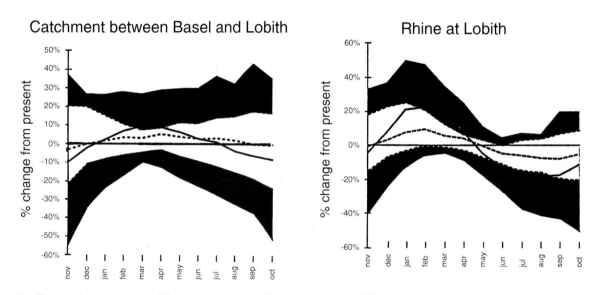

Figure 5.6 Changes in monthly runoff in three parts of the Rhine basin, under different climate change scenarios (Kwadijk and Romans, 1995).

river flow regime types in Nordic countries in warm and cool periods, in an attempt to find temporal analogues for climate change. They found that some regime types remained stable as climate changed but others, particularly transitional types influenced by both snow and rain, changed characteristics.

5.3.3 Eastern Europe

Hydrological regimes in eastern Europe are, like those in the Nordic countries, dominated largely by snowmelt, and the same general conclusions have been drawn in eastern Europe as in other environments with significant snow storage: global warming would lead to reduced spring and summer dis-

charges, and increased flows during winter. Kaczmarek and Krasuski (1991), for example, simulated a decrease of 24% in summer flows in the Varty basin in Poland, and an increase in winter discharges of 21%, from a scenario based on output from the GFDL GCM. Similar results for the Vistula basin were found by Ozga-Zielinska et al. (1994) using a monthly conceptual water balance model. Because winter precipitation would no longer be stored as snow, the timing of the flood season would be brought forward from spring to winter; this has been observed during the late 1980s and early 1990s in Poland.

Gauzer (1994) used a flow forecasting model to simulate the effects of changes in temperature on flow regimes in the

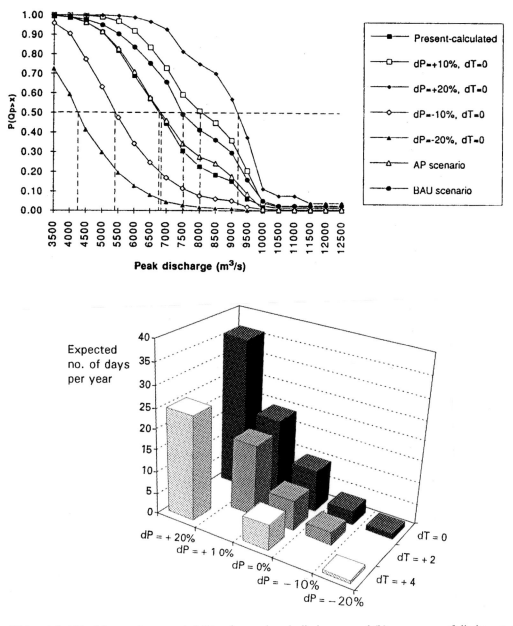

Figure 5.7 The Rhine at Lobith: (a) exceedance probability of annual peak discharges, and (b) occurrence of discharges greater than 5500 m³/s, under various scenarios (Kwadijk and Middelkoop, 1994).

Danube at Nagymaros, Hungary. Figure 5.9 shows the effect of a 3°C increase in temperature on average flows: the snow-induced peaks are reduced, but the peaks due to rainfall alone are not affected.

The Volga basin, with an area of 1.38 million km², is the largest in Europe, and drains much of European Russia. The hydrological regime in the basin is heavily modified. Reservoirs in the basin have a total surface area of 25 000 km², and 2 million ha are under irrigation. By the late 1980s, annual runoff at the mouth of the Volga had been reduced by between 8 and 10% compared with the

1946–65 average, and the increasing use of water within the basin is predicted to result in a total decline of between 14 and 16% by the first decade of the twenty-first century (Shiklomanov, 1989). Superimposed on these human influences will be the effects of climate change. These would have consequences not just for the Volga basin itself, but also for the Caspian Sea, which receives 80% of its inputs from the Volga.

The State Hydrometeorological Institute in St Petersburg has investigated the potential effects of climate change on water in the Volga basin, using three paleoclimatic analogues

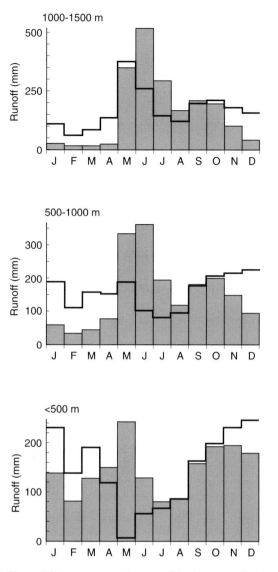

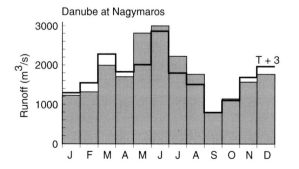

Figure 5.8 Average monthly runoff in three elevation bands in the Vosso catchment, western Norway, under current and changed climates (Sælthun *et al.*, 1990).

Figure 5.9 Average monthly runoff in the Danube at Nagymaros, Hungary, with a temperature increase of 3°C (Gauzer, 1994).

(representing temperature increases of 1, 2 and 3–4°C) and one scenario for changes in temperature and precipitation derived from the GFDL climate model. Each scenario assumes an overall increase in precipitation in the basin, but with a decline in the northern parts. A water balance model, operating on 10-day time steps, was applied to a number of representative sub-basins (Figure 5.10), and changes relative to the 1946–65 baseline period were assessed. Table 5.2 presents the changes in annual runoff (in mm) for 11 sub-basins, under the four scenarios, and Figure 5.11 shows the monthly runoff for the Upper Volga basin under the current climate and the GFDL scenario. Under the 1°C paleoclimatic analogue, there is little change in annual runoff, although some basins do show decreases. Under the other scenarios, runoff increases in all basins and, as Figure 5.11 shows, the spring snowmelt peak tends to be reduced and winter runoff increased.

5.3.4 Southern Europe

Southern Europe has a Mediterranean climate, in which precipitation is concentrated in winter. Indeed, over much of the region there is virtually no rainfall during summer, and annual potential evaporation is higher than annual rainfall. There have been very few studies of changes in hydrological regimes and river flows in southern Europe due to global warming, although considerably more work has been done on desertification problems.

Mimikou and Kouvopoulos (1991) studied changes in runoff volumes in three basins in mountainous central Greece, using a monthly water balance model calibrated to each site and arbitrary change scenarios. Some of the results of the study are shown in Table 5.3. Two of the catchments (Mesohora and Sykia) have a significant snowmelt component, whilst one is largely unaffected by snowfall and snowmelt. The two snow-affected catchments are more arid than the third catchment, and consequently have lower runoff coefficients. They are also more sensitive to changes in precipitation and temperature.

5.3.5 Alpine Europe

Mountainous environments are characterized by snowfall and, in the Alps and parts of Norway, glaciers. Mountain climates show very significant spatial variability – in three dimensions – and are very difficult to infer from the regional climate, so it is much harder to define climate change scenarios from climate model output for mountain environments than for low-lying regions. There have therefore been few

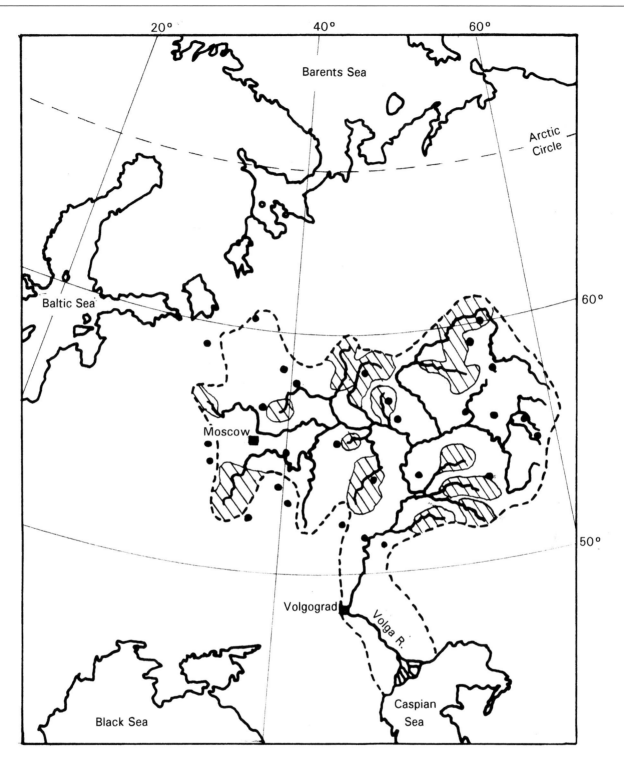

Figure 5.10 The Volga basin: locations of the study sub-basins (shaded) and meteorological stations (•).

studies concerned with changes in alpine catchments in Europe.

Collins (1989) showed that if the proportion of the catchment covered by glaciers is low, then runoff variability in alpine basins is determined by variability in precipitation. If, however, the proportion covered by glaciers is high, then variations in river discharge are caused more by variations in summer energy inputs. Two adjacent mountainous catchments – one with a glacier and one without – might therefore respond very differently to a given climate change.

Table 5.2. *Changes in annual runoff (in mm) in sub-basins of the Volga basin.*

Sub-basin	Area (km²)	Runoff (mm/yr)	Scenario 1°C	Scenario 2°C	Scenario 3–4°C	Scenario GFDL
Kostroma	8 870	242	8.5	21.7	61.5	58.1
Vetluga	27 500	226	2.1	29.5	61.9	127.3
Upper Kama	46 300	258	19.5	31.2	65.2	92.0
Chepsta	18 900	208	24.1	21.4	56.8	71.1
Kokshaga	2 140	188	− 14.7	14.2	49.6	30.4
Sok	4 730	119	− 13.9	9.3	39.4	90.3
Upper Volga	21 100	233	11.7	11.0	84.4	24.5
Upper Oka	54 900	168	− 9.5	12.1	55.8	25.8
Sura	50 100	111	− 6.5	29.1	59.0	94.8
Irgiz	18 200	39	− 0.4	15.4	39.4	52.0
Samara	28 800	64	0.6	15.3	44.3	63.1

Table 5.3. *Percentage changes in annual and seasonal runoff, Greece, with a 2°C increase in temperature (Mimikou and Kouvopoulos, 1991).*

Catchment	Precipitation down 10% Annual	Precipitation down 10% Winter	Precipitation down 10% Summer	Precipitation up 10% Annual	Precipitation up 10% Winter	Precipitation up 10% Summer
Mesohora	− 14	− 19	− 9	15	17	10
Sykia	− 14	− 18	− 10	15	16	11
Pyli	− 12	− 13	− 11	12	14	11

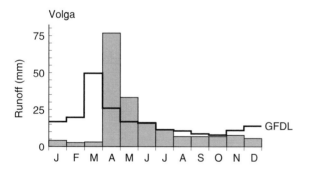

Figure 5.11 Average monthly runoff in the upper Volga basin, under current and changed climates.

Collins (1989) also noted that as glacier area reduces due to glacier melt, then the hydrological regime of the catchment would become more dependent on precipitation variability.

Bultot *et al.* (1994) studied snow accumulation and meltwater flooding in a high alpine basin in Switzerland. They found that snow accumulation was very sensitive to temperature variations, and that global warming would lead to significant reductions in accumulation and meltwater floods, and an increasing susceptibility to rain-generated floods.

5.3.6 Current activities

Undoubtedly many studies of effects of climate change on hydrological regimes are being undertaken in many catchments in Europe, but little information on these projects is available. The European Union, however, is funding a coordinated investigation into the effects of climate change on hydrological regimes in Europe (Project EV5V-CT93-0293: Arnell, 1995), which is looking at the European, regional and catchment scales, using a consistent baseline period (1961–90) and consistent climate change scenarios derived from climate model output. Different hydrological models will be applied at each scale and in each catchment, but a consistent impact assessment methodology is being adopted. The consortium includes 16 institutions from western and eastern Europe, and the project was scheduled to finish in 1996.

Another EU-funded project, GRACE (Groundwater Resources and Climate Change Effects; EV5V-CT94-0471) is investigating the sensitivity of carbonate aquifers to changes in the recharge rate. The focus is on changes in aquifer properties, including hydraulic parameters and carbonate dissolution/precipitation rates (and hence water quality).

5.4 INTERNATIONAL PROJECTS AND EXPERIMENTS

Several meso-scale field experiments have been and are being undertaken in Europe, primarily aimed at investigating large-scale fluxes of water and energy in order to contribute to the improvement of climate simulation models. These experiments therefore contribute *indirectly* to the understanding of the effects of climatic variability and change on hydrological characteristics in Europe. One project, however, is concerned directly with the hydrological implications of climatic variability, and this will be summarized first.

5.4.1 FRIEND

The FRIEND (Flow Regimes from International Experimental and Network Data) project has made a recognized contribution to UNESCO's third, fourth and fifth International Hydrological Programmes (Gustard *et al.*, 1989; Gustard, 1993; Roald *et al.*, 1989; Seuna *et al.*, 1994). This collaborative international project, which began in 1985, is concerned with the analysis of regional hydrological regimes and variability in Europe. Most studies have been undertaken in northern and western Europe, but during the 1990s the study area was expanded to include eastern Europe, and studies are also beginning in southern and Mediterranean Europe (in the Alpine and Mediterranean Hydrology project, AHMY).

The FRIEND project is organized into research themes, underpinned by an international database of river flows and physical catchment characteristics. The European Water Archive currently holds data from over 2500 catchments in northern, western and eastern Europe, mostly with catchment areas of less than $1000 \, km^2$. One of the FRIEND research themes is centred around the analysis of large-scale hydrological variability in Europe (Arnell *et al.*, 1993). The project has explored the definition and variability of hydrological regimes (Krasovskaia and Gottschalk, 1992), and has mapped annual runoff across Europe. Initial descriptive analyses of variability in European hydrological regimes (Arnell, 1994) has shown the strong spatial correlation in patterns of variability over time (Figure 5.1), and work is currently under way to relate these patterns to atmospheric and sea surface temperature anomalies.

5.4.2 HAPEX-MOBILHY

The Hydrological Atmospheric Pilot Experiment-Modélisation du Bilan Hydrique (HAPEX-MOBILHY)

experiment was one of the first major international meso-scale field experiments. It took place between 1986 and 1988 over a $100 \times 100 \, km$ area in southwest France (André *et al.*, 1988). As in all meso-scale field experiments, HAPEX-MOBILHY included a dense network of ground measurements and coordinated measurements of surface and atmospheric fluxes from ground and airborne instruments during a few intense observing periods. Measurements were made over a range of vegetation types, including forest and cornfields. Results from the study have been used to improve the representation of the effects of forest cover on meso-scale rainfall, for example (Blyth *et al.*, 1994), and have shown the significance of the forest edge for the generation of rainfall.

5.4.3 NOPEX

The Nordic Hydrological Experiment (NOPEX) is a meso-scale land surface experiment being conducted over a $100 \times 100 \, km$ area near Uppsala, Sweden (Halldin *et al.*, 1995). Its specific aim is to investigate fluxes of energy, momentum, water and CO_2 at local to regional scales, using both measuring and modelling. The landscape in the NOPEX region is a mosaic of small fields and marshes, with limited topography and impeded drainage; the depth of the groundwater level has a major influence on hydrological regimes and evaporation. The first intensive observations were made during 1994 and 1995, and intensive field campaigns were scheduled until 1996. The project will enable improved simulations of land surface processes in cool, patchy landscapes.

5.4.4 BALTEX

The Baltic Experiment (BALTEX) is primarily concerned with measuring and modelling the fluxes into and out of the Baltic Sea. It began in 1994, and a major component of the project involves the refinement of a coupled regional climate and ocean model. The major research phase will take place in the period 1997–2001 (BALTEX, 1995).

5.4.5 MEDALUS and EFEDA

The aims of the EU-funded project Mediterranean Desertification and Land Use (MEDALUS) are to gain an understanding of the process of desertification, to model desertification and its impacts, and to identify possible mechanisms for mitigating these effects. The study involves the analysis of contemporary, past and future climates in the

Mediterranean, and measurement of the physical processes of desertification at seven main field sites around the Mediterranean. The project is also developing models to simulate runoff, soil erosion and sediment yield, and is applying remote sensing and GIS technologies to identify dynamic response units.

The European Field Experiment in Desertification-Threatened Areas (EFEDA) is also concerned with deforestation, but is concentrating on the interaction between land surfaces and atmospheric processes. It is, as the name suggests, a field experiment, located in southwest Spain. It began in 1991 (Bolle et al., 1993), and includes intensive campaigns of detailed measurements, from land, air and space, of land surface characteristics and fluxes over three different types of land cover.

5.5 CONCLUSIONS

This chapter has reviewed many of the studies of the effects of climate change on hydrological regimes in Europe. It is, however, very difficult to compare the studies and results, because they have used different methodologies and, most importantly, different climate change scenarios. Some studies have used arbitrary changes in climatic inputs, some have used temporal analogues, and others have used scenarios based on paleoclimatic reconstructions. Scenarios based on climate model output have used different climate models, and different ways of applying scenarios at the catchment scale. Most studies have used conceptual water balance models, but the time steps at which these models operate vary,

and a few studies have used empirical relationships between climate and runoff. Clearly there is a need for a consistent study using similar scenarios across the whole of Europe, as is currently being undertaken with EU funding.

The most general conclusions so far are that global warming may affect significantly the timing of river flows in catchments controlled by snowmelt, and that changes in the seasonal distribution of river flows may be more significant than changes in annual totals. In many catchments in western Europe (in the UK, Belgium and Switzerland), the climate change scenarios used imply increased concentrations of runoff during winter.

Although there have been many studies in Europe, there are still two important gaps. First, there is very little information on possible changes in groundwater recharge, which is a major source of public supply. Second, there have been no assessments of the relative importance of climate change and land use change; at the local scale these latter changes may be considerably more important than climatic change.

There have been a few intensive meso-scale field experiments aimed at understanding the linkages between the land surface and climate, and these, together with similar experiments being undertaken in different parts of the world, will lead in the medium term to improvements in the simulation of current and future climate. Only once the complexities of European climate can be adequately modelled will it be possible to define credible change scenarios *at the catchment scale*. It is also important to understand more fully the nature of the links between climatic variations and hydrological anomalies in Europe, and the UNESCO FRIEND project provides the ideal vehicle.

6 Assessment of the impacts of climate variability and change on the hydrology of Africa

J. SIRCOULON, T. LEBEL and N. W. ARNELL

6.1 INTRODUCTION

Africa, with an area of 30.5 million km^2, is the Earth's second largest continent, and presents the most extreme climatic contrasts. The Sahara is the largest and most extreme desert; Mount Cameroon is one of the rainiest places in the world; and the Congo River, which accounts for 30% of the runoff of the entire continent, is the world's second-ranking river in terms of discharge. This great diversity of climates and hydrological structures has been accentuated by very sharp variations over time and space.

The water resources of Africa are very unevenly distributed. The arid and semi-arid zones, which account for 60% of the continent's land area, produce only 5% of the total runoff (Margat, 1991). Although this very unbalanced situation is attenuated by such major rivers as the Niger, Nile and Zambezi, it means that some countries are dependent on their neighbours for their water – a situation that may in future lead to disputes over the allocation of these resources.

In recent geological time there have been marked climatic changes (as witnessed by fossil drainage networks and rock paintings), with sudden fluctuations in rainfall levels. In the Sahel countries the drought that has persisted since the late 1960s has greatly reduced the discharge rates of the main tropical rivers and has led to the drying up of Lake Chad almost completely (Sircoulon, 1987). In contrast, however, exceptionally heavy rainfall can also occur – as it did in 1961–64 in the Upper Nile and Congo basins – leading to an abrupt rise in the water levels of the East African lakes (Piper et al., 1986) and increasing the inflows from the White Nile to the Main Nile.

When studying the climate, its evolution and its impact on hydrological regimes, a number of constraints have to be taken into account:

● Climate data are scarce, and there are few long-term records for either precipitation (Nicholson et al., 1988) or runoff (Sircoulon, 1976). Moreover, in recent decades the hydrometric networks have deteriorated (WMO, 1989b; Water Assessment, 1993), with regard to both the number of usable stations and the quality of observations.

● General circulation models (GCMs), although still not widely used in Africa, broadly agree that temperatures will rise moderately (by 1–2°C) with a doubling of CO_2 levels. However, the impact of such warming on precipitation is by no means clear. The GCM model results diverge considerably, especially for the arid zone (Mabbutt, 1989; Santer et al., 1990; IPCC, 1990a). Very few model-based studies have been made of the impacts of climate change on runoff in major drainage basins such as the Nile (Conway, 1993, Hulme et al., 1994); some that do exist are not available at present (Zambezi study).

Studies such as the Global Energy and Water Cycle Experiment (GEWEX), which have been designed to improve our knowledge of the exchange processes between soil, vegetation and the atmosphere at a regional level (the mesh size of GCMs), are still in their infancy. The only major operation of this kind so far has been HAPEX–Sahel (1992); the data from this study have not yet been fully exploited, but the innovative nature of the experiment, the first results and the prospects it opens up justify the emphasis in Section 6.2.3.

This chapter describes the situations in West and Central Africa (Section 6.2), East and Northeast Africa (Section 6.3). Conclusions and recommendations for Africa as a whole are presented in Section 6.4.

6.2 WEST AND CENTRAL AFRICA

This section deals with the recent climate fluctuations observed on the Sahel–Sahara fringe, the lessons that can be learned from the present drought, and the international research programmes.

6.2.1 Recent climate fluctuations at the Sahel–Sahara fringe

A glance at the climate changes that have occurred since the last (*Würm*) glacial maximum and during the present interglacial (i.e. in the past 20 000 years) shows that temperatures and precipitation have varied considerably, leading to major environmental changes, as reflected in the type and abundance of vegetation and the behaviour of lakes and watercourses.

Paleoclimatic background (20 000–1000 BP)

Studies of paleoenvironments combine various approaches: analyses of pollen (Maley, 1981), Ostracoda, diatoms (Servant-Vildary, 1978) and plant remains, paleohydrology, reconstructions of lake level fluctuations (Street and Grove, 1979; Nicholson, 1980), runoff regimes and ancient drainage networks (Dubief, 1953), geomorphology and isotopic geochemistry. Archeological research (Roset, 1987) has revealed that lakeside settlements were established during periods of abundance, whereas civilizations declined during periods of drought (Bell, 1971). The following periods can be identified:

- *20 000–12 000 BP.* A markedly dry period (Rognon, 1989), when continental temperatures must have been 3–4°C lower than today, with persistent winds. Advancing sand dunes choked major rivers such as the Senegal, Niger and Nile, and many lakes dried up.
- *12 000–7500 BP.* Rainfall increased, peaking around 9000 BP (the Holocene pluvial). This made it possible for the development of Neolithic settlements in the Sahel and Sahara (Roset, 1987). Flood flows in African watercourses were considerable (Evans, 1994; Saïd, 1994); and lakes such as Chad and Victoria reached their highest levels. Rainfall seems to have been fairly well distributed over the year (probably 30% more abundant than the 1931–60 norm; see COHMAP, 1988), and continental temperatures clearly allowed a favourable biotope with abundant vegetation.
- *7500–2500 BP.* After a dry period (7500–6000 BP), during which the Saharan lakes shrank, a new but less pronounced wet phase set in between 6000 and 2500 BP, with considerable fluctuations in lake levels throughout Africa. Rainfall patterns seem to have changed, with a long dry season separated by very heavy

rains, and temperatures were higher than those of today.
- *2500–1000 BP.* The climate has become drier, the rainy season shorter and evaporation greater.

One of the main points to emerge from modern accurate dating methods is that there have been very rapid shifts (i.e. within a century or so) to wetter conditions (e.g. a spectacular northward advance of vegetation around 9000 BP) or drier conditions (around 2500 BP).

Historical period (1000–100 BP)

There are a number of ways of assessing climate change during this period (Maley, 1981; Nicholson, 1981a), but the different methods need to be carefully compared and correlated. These include studies of oral traditions, historical documents and Arabic chronicles; radiocarbon dating of core samples, tracing the history of flood levels of rivers such as the Nile (Jarvis, 1936; Popper, 1951); historical inquiries (Olivry and Chastanet, 1986) and explorers' tales. Taken together, these sources paint the following approximate picture for the regions immediately south of the Sahara:

- seventh to the twelfth centuries: wetter conditions than today;
- thirteenth century: drought and renewed fierce winds in the Sahara;
- fourteenth to the end of the sixteenth century: a marked climatic improvement, but several periods of accentuated drought and famine around 1420–60 and 1550, with Lake Chad at low levels (Maley, 1981);
- seventeenth and eighteenth centuries: the Saharan zone became drier while the tropical regions received more rain (Lake Chad at a high level). Severe droughts seem to have occurred in about 1680 and 1750 (Nicholson, 1980, 1981a);
- nineteenth century: a rainfall peak in the second half of the century (Nicholson, 1980), corroborated by travellers' accounts (Barth, 1860; Rohlfs, 1874; Nachtigal, 1881) and high Nile discharges (Hurst, 1965).

Instrumental period (100 BP to today)

Temperature

From temperature records kept since 1895, Jones and Briffa (1992) have shown that temperatures have risen by 0.53°C over the last century, with two periods of sharper increases, 1910–40 and 1980–90. These records correlate well with trends observed at the global scale.

Rainfall

Despite the low density of rainfall networks and the small number of long-term stations (i.e. in operation since the turn of the century), the regional indices (Lamb, 1985; Nicholson *et al.*, 1988) show that there have been considerable fluctuations, with two wet periods and three dry periods (around 1913, 1940 and since the late 1960s).

The present drought (although 1994 was the wettest year since 1964, according to the NOAA Climate Analysis Center) has shown two peaks of intensity, in 1972–73 and 1983–84. This has been by far the most severe, extensive and persistent drought of the century. The 200–600 mm isohyets have shifted southward by several hundred km in the Sahel zone (Albergel *et al.*, 1985), while the rainfall averages for 1961–90 were 20–40% lower than those for 1931–60 (Hulme, 1992). Table 6.1 shows the trends in average annual precipitation since 1951 for some of the major tropical and equatorial river basins. The table shows that rainfall has also declined in humid West Africa. Although the differences are smaller, they mask considerable spatial and temporal disparities.

Runoff

Many studies have been made of the changes in watercourses during the recent drought (changes in regime, sharp drop in flood discharges, accentuation of low-water levels, more frequent drying up) and of the increasing scarcity of water resources (e.g. Sircoulon, 1976, 1987, 1990; Sutcliffe and Knott, 1987; Olivry *et al.*, 1993; Mahé, 1993). Mahé (1993), for example, showed that there has been a considerable drop in annual inflows to the Atlantic from African rivers over the past 30 years, the decade 1981–90 being the most severely affected (see Table 6.2). The yields of tropical rivers reaching the Sahelian zone have also fallen (Sircoulon, 1990); up to and including 1969, the average yield from the Senegal, Niger and Chari rivers was 136 km^3 per year. In the period 1970–88, it fell to 79 km^3 (a deficit of 43%), and in 1984 it was no more than 36 km^3 (a shortfall of 74%).

This persistent rainfall deficit has naturally had a major impact on lake environments, the shrinking of Lake Chad being the most spectacular example. In 1962–63 the lake was at its highest level since the beginning of the century, with an area of 23 500 km^2 and an estimated volume of 105 km^3, but then the level dropped until, in 1973, it split into two separate basins. Over the last ten years the dry season area of Lake Chad has shrunk to about 1500 km^2, and the northern basin sometimes remains dry for several years in succession (Pouyaud and Colombani, 1989). Studies have shown that two successive years of abundant rain would be needed to fill

it completely (Lemoalle, 1989); although the rains returned in 1994, river inflows to the lake still showed a 30% deficit.

6.2.2 Lessons from the present drought

The prolonged drought in the Sahel, the effects of which have reached as far as the equatorial zone, at least in an attenuated form, has provided important quantitative data on trends in different features of the Sahelian, tropical and equatorial hydrological regimes (annual runoff, flood discharges, low water discharges, duration of low water levels or dry streambed periods). Studies of the total impact of the Sahelian drought on the environment are also valuable in that they show that models simulating climate change must take into account a number of factors:

- the lasting effects of desertification (Hare, 1983) on the state of various ground surfaces, since the soil tends to become impermeable (Casenave and Valentin, 1990);
- changes in runoff conditions in basins subjected to the combined impacts of climate and human activities, such as cultivation and spreading soil erosion (Albergel and Gioda, 1986; Albergel, 1987);
- changes in the behaviour of groundwater reservoirs (Olivry *et al.*, 1993) in the large river basins, with the persistent drought introducing a memory effect on aquifers; consequently there is a progressive increase in the depletion coefficient;
- the effects of scale (land area), since the present runoff conditions in Sahelian drainage basins are a function of area (Pouyaud, 1987; see below).

In this connection, it should be borne in mind that rainfall deficits lead to even more severe runoff deficits (such as occurred in 1984). Even in humid Africa, rainfall deficits of a few percent but of very long duration can lead to a marked depletion of water resources. In 1981–90, water resources fell by 65 km^3 a year in dry Africa and 365 km^3 in humid Africa. A deficit in groundwater reserves has a delayed impact on the major river basins, especially in the tropical zone. Even when rainfall increases, as it did in 1994, for example, the prolonged accumulation of groundwater deficit means that runoff increases far more slowly. On the other hand, the increasing impermeability of soils in the Sahelian zone explains why the maximum flood discharges in many Sahelian watercourses remain high and the total annual runoff remains stable, despite the markedly reduced precipitation (hence the higher runoff coefficients). This needs to be remembered particularly in the design of water engineering structures. The larger the basin, however, the less marked this effect will be, due to losses by evaporation and hydro-

Table 6.1. *Average annual precipitation in the major tropical and equatorial river basins in Africa (in mm/yr) (Olivry et al., 1993).*

Basin	Area (km²)	1951–60	1961–70	1971–80	1981–90	% Difference $\left(\dfrac{1981/90-1951/60}{1981/90}\right)$
Senegal	218 000	1071	985	843	766	− 28
Niger	120 000	1649	1527	1403	1315	− 20
Gambia	42 000	1365	1226	1049	991	− 27
Sanaga	131 200	1924	1867	1800	1722	− 10
Oubangui	48 800	1578	1573	1486	1515	− 4
Ogooue	203 000	1792	1839	1757	1776	−1
Zaire	3 500 000	1511	1467	1466	1440	−5

Table 6.2. *Annual inflows to the Atlantic from African rivers, 1951–90 (in km²/yr) (Mahé, 1993).*

Hydrological region	Mean 1951–90	Mean 1981–90	% Difference $\left(\dfrac{1981/90-1951/90}{1981/90}\right)$
Humid Africa (Zaire to Guinea)	2350	1985	− 16
Dry Africa (Dakar, Gambia, Niger)	235	179	− 24
Total	2585	2164	− 17

graphic degeneration. In basins of more than 20 000 km², watercourses in the Sahel show the same deficit patterns as tropical rivers.

The most recent 25-year drought in the Sahel has had a marked impact on the hydrology, and consequently on the living conditions of the inhabitants of the region. The unusual length of the drought, at least with respect to available records (which unfortunately go back little further than the beginning of the century), has raised a number of questions as to the significance of this climatic episode and its possible causes (Hubert and Carbonnel, 1987; Nicholson, 1981b).

In 1975, Charney suggested that the persistent drought could be explained, at least in part, as being due to the environmental impact of overgrazing. Overgrazing reduces the amount of vegetation and raises the albedo. This in turn aggravates and extends the atmospheric subsidence over the Sahara, preventing rain in the Sahel and furthering the disappearance of vegetation. The consequence of this positive feedback loop is an acceleration in the advance of the desert. The conflicting results of simulations, using various GCMs, make it impossible to draw firm conclusions on the validity of a large-scale Charney mechanism. The first simulations carried out by Charney suggested a southward dis-

placement of the Intertropical Convergence Zone (ITCZ) by about 4° and an increase in rainfall at latitudes 10°–14°N. However, this trend has not been confirmed by rainfall data recorded over the last 40 years. A variety of other mechanisms have been invoked to explain the persistent Sahelian drought, including the role of reduced soil moisture (Sud and Molod, 1988; Rowell and Blondin, 1990). Changes in the sea surface temperature (SST) in the surrounding oceans have also been proposed as a possible cause by Lamb (1978), Folland *et al.* (1986) and Palmer (1986).

The fact is that the GCMs are not yet capable of determining, with any reasonable degree of certainty, the consequences of environmental change or of the greenhouse effect on rainfall in Africa. Some models predict an overall decline in rainfall for the whole of West Africa, whereas others single out some areas in which rainfall will be unchanged, or will increase, and others in which it will decrease (Laval and Picon, 1986). Nevertheless, recent studies (Xue and Shukla, 1993) have simulated a complete set of 'normal' and 'drought' conditions, representing a relatively new approach. So far, in GCM simulations, one surface parameter after another has been varied in attempts to determine the sensitivity of the model to changes in a given parameter. The simulations of Xue and Shukla (1993) have demonstrated the crucial role played by vegetation in the development of possible feedback effects. Despite this, the scope of their conclusions is limited by the lack of real data required to specify atmospheric and surface conditions in the Sahel.

To some extent, these uncertainties reflect the shortcomings of GCMs, and are related to insufficient resolution or oversimplified parameterization. For tropical regions, however, the uncertainties also reflect how little we know about the processes that govern the interactions between the surfaces of continents and the atmosphere. Our poor knowledge is due to the lack of specific studies and the inadequacy of the instrumentation that is generally available, particularly in Africa. It should also be stressed that it has not been proven that the present Sahelian drought, which has lasted since 1968, is a manifestation of long-term changes in the climate of the region. The available series of records are not sufficiently long to support this theory.

There are many urgent reasons – both local (such as the need for economic development, and sometimes the survival of populations) and global – to find a remedy to a situation of this nature. The Intertropical Convergence Zone represents the heat source driving atmospheric circulation. If we do not improve our knowledge of this heat source, our understanding of the planet's climate and the scenarios used to predict future trends will remain imperfect. Following the success of the Hydrologic–Atmospheric Pilot Experiment–Modélisation du Bilan Hydrique (HAPEX-MOBILHY), which has been designed to study in detail the processes that control interactions between the continents and the atmosphere in a temperate zone (André et al., 1986), conducted in southwest France (1985–86), it was decided to mount a comparable project for the Sahel, under the name of HAPEX-Sahel. The choice of the Sahel was influenced by two considerations. First, it is an area in which the balance between the population and the environment is particularly fragile. Second, the Sahelian environment comprises a limited number of vegetative units, reputedly fairly homogeneous, which may be expected to favour the spatial aggregation of the processes and consequently their representation in a GCM. HAPEX-Sahel was preceded by a smaller-scale study restricted to rainfall (the Estimation of Precipitation by Satellites in Niger, EPSAT-Niger). These two projects, the experimental parts of which are now complete, have yielded data on all the processes affecting the continent–atmosphere interface in the Sahel.

6.2.3 International research programmes

(1) The EPSAT-Niger experiment

The EPSAT-Niger experiment (Lebel et al., 1992) was started in 1988 by a group of French laboratories (ORSTOM; the Laboratoire d'Aérologie, Toulouse; the Laboratoire de Météorologie Dynamique, Palaiseau; and the CNRS-URA 1367, Fontainebleau), in collaboration with Direction de la Météorologie du Niger (DMN) and Niamey University. Its main objectives were: (1) to improve rainfall estimation algorithms for the Sahel, combining the available data sources (rain gauges, radar and satellites); (2) to characterize and model rainfall fluctuations from the local level to the meso-scale, providing information of use at the climatic level; and (3) to study the consequences of these fluctuations on the hydrology of the Sahel, on the basis of additional measurements set up for HAPEX-Sahel.

The experimental system comprised 100 automatic recording rain gauges and a digital C-band weather radar system, covering an area of one square degree ($2°–3°$ E; $13°–14°$ N), which also served as the basis for HAPEX-Sahel. The rain gauges were installed to form a network with a 12.5×12.5 km mesh size, which is equivalent to about 3×3 Meteosat pixels. At the centre of this network, a target area of 20×20 km was equipped with about 20 instruments in order to study the variability at a smaller scale (Figure 6.1).

After two years of preparatory work in 1988 and 1989, the complete system was in operation from 1990 to 1993. A climate watch with about 40 devices began in 1994. If possible, the system will be maintained in order to collect climatological data for the area over a further ten-year period. It will also serve as a means of validating the data provided by the Tropical Rainfall Measurement Mission (TRMM; Simpson et al., 1988) and Tropiques (Desbois, 1994) satellite missions.

Until recently, the only data suitable for detailed analyses of the structure of the rainfall fields of tropical Africa were provided by GATE (GARP Atlantic Tropical Experiment), and they concerned the oceans. On the African continent, the data from the dense rain gauge network at Ouagadougou (1987–88) has made it possible to test the fractal approach to modelling rainfall occurrences, but the absence of data for small time steps (recording rain gauges or radar) limits its value. The EPSAT-Niger data will fill many of the gaps left by GATE and the Ouagadougou network. They concern a continental region and combine ground and radar data, covering a period of several years, which should make it possible to obtain meaningful climatological representations for convective structures.

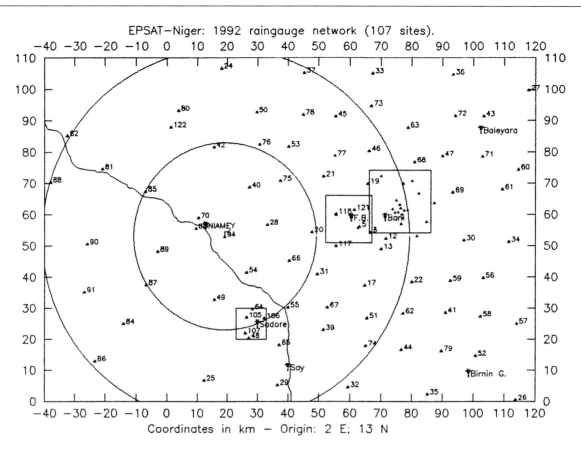

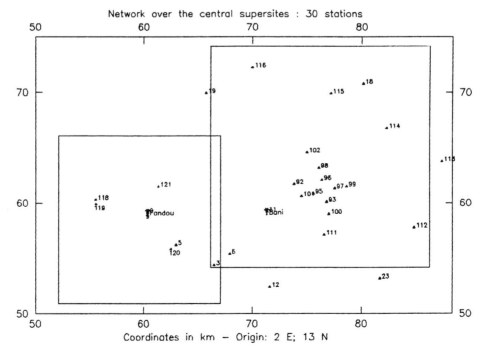

Figure 6.1 The EPSAT-Niger rain gauge network in 1992.

Perhaps one of the most important first results of this experiment has been the accuracy of average rainfall estimates for a given surface area. Using the recording rain gauges on a network with a 12.5 × 12.5 km mesh size, the accumulated rainfall for the duration of an event can be calculated with an average error of less than 15% for surface areas ranging from 1000 to 10 000 km^2. The error increases to 30% for areas of about 100 km^2. An important result at this initial stage in the data analysis is that the error in the ten-day, monthly and annual rainfall totals is about the same as that for the totals per event. The recording rain gauge data can also be used to study the space–time distribution for time scales equal to or smaller than the rainfall event, thus providing climatological information on the number of events observed during a rainy season, and, above all, the average rainfall for these events. The distribution of the event totals, at a given point, seems to be almost exponential, remaining very stable from one season to the next, with an average and a standard deviation both equal to about 15 mm. The space–time distribution, on the other hand, is very intermittent. More than half of the seasonal total, at a given point, can fall in less than 5 hours, and disparities as high as 100% have been observed for points only 10 km apart. Differences of this sort, which have so far been acknowledged when working at the event scale, have never been measured for annual rainfall. This is easy enough to understand, given that about 50% of the annual total falls with an intensity exceeding 35 mm/hour, and that 35% exceeding 50 mm/hour (Lebel *et al.*, 1995).

The high spatial variability at all time scales means that the convective structures that make up the Sahelian rainfall systems must be taken into account when modelling rainfall fields, even if several events are combined. Radar data may prove useful in this respect, provided that they are first properly calibrated. Considerable methodological work has been done to this end. For instance, it was shown that the C-band radar signals were significantly attenuated by squall lines, and required careful correction.

The team processing the EPSAT-Niger data for hydrological purposes will now concentrate on developing a model for the rainfall fields taking into account the dynamics specific to the Sahelian meso-scale convective systems. To achieve this, a more detailed climatic description than that currently available will be extracted from ground data. Radar data will also be used, but for this type of application will require careful calibration. At the same time, work will continue to obtain a clearer understanding of the implications of the intermittence of rainfall fields for satellite measurements, associated or not with measurements on the ground (recording rain gauges and/or radar). In particular, the EPSAT-Niger data could be used to simulate the consequences of the sampling used for the TRMM and TROPIQUES satellite missions and to validate the estimation algorithms to be used in this framework.

(2) *The HAPEX-Sahel experiment*

The HAPEX-Sahel programme (Goutorbe *et al.*, 1994) was started as part of the GEWEX initiative, under the World Climate Research Programme, to improve our knowledge of the fluxes of heat, water vapour, and momentum over the Sahel, which represents the largest continental area subjected to a semi-arid tropical climate. More specifically, the programme aims to improve the hydrological representation of semi-arid zones in GCMs and to obtain a clearer idea of the main factors behind the desertification of the Sahel over the last 20 years. Remote sensing, the only technique that allows large-scale observations, is playing an important role, and specific research has focused on interpreting satellite data, but the significance of the results derived from remote sensing in this area is seriously affected by the presence of numerous aerosols in the atmosphere.

Specific features of the Sahelian environment

The climate of continental West Africa is influenced by the interaction of the air masses over large, homogeneous areas of sea and land. The two air masses meet in a zone which is the ascending branch of the Hadley cell, also referred to as the ITCZ. The tropical continental air mass originating from the Sahara desert contains warm, dry air that is generally hazy. It slopes upwards to the south over the warm, humid tropical maritime air originating over the Atlantic, which forms a wedge under the continental air mass. The seasonal migration of the ITCZ is fundamentally important in our understanding of the climate of West Africa.

The Sahelian region comprises the semi-arid vegetation belt south of the Sahara desert and north of the dry forest zone. The duration of the rainy season decreases from about five months in the south of the Sahel (12° N) to three months in the north (18° N). The annual rainfall is closely related to the duration of the rainy season, varying from 800 mm in the south to only 200 mm in the north, following a regular gradient of 1 mm/km (averages from 1950–89; Lebel *et al.*, 1992).

The highly variable and unreliable rainfall governs the growth and distribution of vegetation. Total rainfall everywhere is less than the potential evaporation, which is of the order of 2000 mm/year. This results in extremely sparse vegetation with large areas of bare soil. The interlacing of bare soil and vegetation creates a large variability of surface conditions at the 0.01–1 km^2 scale. However, the somewhat uniform geological and pedological structures result in a relatively homogeneous regional landscape.

Another major aspect of the Sahelian environment is the marked degradation of drainage networks, as a result of highly intermittent streamflow characteristics. Sahelian hydrology is therefore primarily the hydrology of endorheic (inward draining) areas, the surface areas of which rarely exceed a few km^2. This results in the formation of pools at the beginning of the rainy season, which dry out 2–3 months after the last rainfall. The areas that do not drain externally do not behave as watersheds in the classical sense either; runoff dissipates rapidly as the slopes weaken. It is therefore generally impossible to rely on measured streamflow at the watershed outlet to provide spatially integrated values of the residual (evaporation) term in the water balance. Emphasis must therefore be placed on meteorological measurement techniques to obtain the (spatially) integrated evaporation.

Experimental strategy

In order to bridge the disparity between the scale of a GCM mesh (squares with sides 100–200 km) and the scale of the surface process fluctuations (hydrology, vegetation cover, etc.), the approach adopted for HAPEX-Sahel was based on both spatial integration using remote sensing data, and on modelling of the atmosphere at sub-mesh scales (meso-scale models with a resolution of 10 × 10 km) and of the hydrology (watershed models). The area covered by the instruments is one square degree, as initially selected for EPSAT-Niger, which represents an area of 110 × 110 km (an order of magnitude larger than a GCM mesh) containing representative samples of the vegetative units of the region.

An hierarchical system with three observation scales was defined. First, the largest area of 12 000 km^2 was used to model responses on the scale of a GCM mesh. We expect to find a state of equilibrium with respect to the exchanges. This large area was covered by the EPSAT-Niger network of rain gauges, a network of 12 automatic weather stations and a network of over 200 points at which the Continental Terminal (CT3) upper aquifer is monitored.

Second, within this area, three 15–20 km^2 super-sites and two secondary sites were selected, corresponding to the definition of meso-scale models, at which most of the ground measurements were made. Sites of this size contain sufficient diversity for them to be regarded as representative of surface conditions. At this scale it is also possible to superimpose field, satellite and aircraft measurements. The locations of the super-sites (Figure 6.2) were chosen to sample the north–south rainfall gradient, which represents a difference

of a little over 100 mm for the annual average between latitudes 13° and 14°, or 1 mm/km. Finally, extensive measurements were made at the plot level. At each super-site, three or four adjacent plots corresponding to the dominant vegetation types (millet fields, fallow and tiger bush), were selected to establish detailed balances at the ground–atmosphere interface for momentum, radiation, energy, water and, to a certain extent, carbon. Satellite and airborne remote sensing resources and measurements taken in the atmospheric boundary layer (radio-sounding and airborne flux measurements) link the dispersed sites.

The measurements were made over three time scales: First, during the long-term monitoring period (1991–93) the focus was on vegetation and hydrology studies. Second, aircraft measurements were concentrated in an Intensive Observation Period (IOP), which ran from 17 August to 8 October 1992. The IOP was timed to capture the transition from the rainy to the dry season, during which both soil moisture and vegetation changes combine with changing meteorological conditions to produce a radical transformation in the surface energy balance of the region. During the IOP several aircraft were involved in both remote sensing and flux measurement missions. The long-term monitoring period allows investigation of hydrological processes operating on seasonal or longer time scales, such as soil moisture and runoff, while the relatively short IOP provides data on processes operating on shorter time scales, such as surface fluxes of heat and water vapour.

Third, during the 1992 rainy season (June–October) all the ground instruments were set up at the super-sites and at the plot level. This meant that a large number of local studies were carried out over a period much longer than the IOP itself. The use of 'long-term' to qualify the period 1991–93 may appear slightly misleading, compared with the time required to sample accurately the inter-annual variability of the Sahelian climate. The relative brevity of operations for the HAPEX-Sahel system was due to the difficulties involved in supporting research teams who were unused to working in Africa and to operating and maintaining sophisticated equipment in somewhat precarious conditions for any length of time. Most of the long-term study work was conducted by teams from ORSTOM who were more accustomed to working for long periods in tropical countries, where their programmes are based.

Measurements

HAPEX-Sahel consisted of 66 investigations (for details, see Goutorbe *et al.,* 1992), using either specific data or data obtained in other studies. They were divided into four groups: hydrology, meteorology and boundary layer, surface

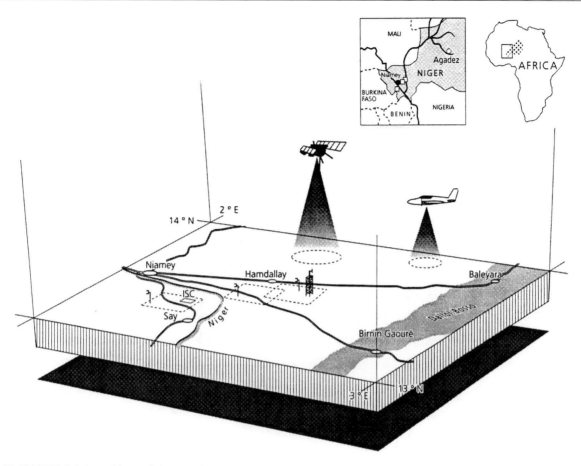

Figure 6.2 HAPEX-Sahel: positions of the experimental square and the three super-sites.

fluxes, and remote sensing, to facilitate their understanding, in scientific terms, and their coordination in the field.

Almost all of the planned experimental programme was completed, despite difficult weather conditions and logistic problems. The long-term monitoring of rainfall and fluctuations in the level of the upper aquifer covered an area of one square degree and lasted much longer than was initially planned, because a smaller but still representative system (about 40 recording rain gauges and over 100 access points for monitoring the aquifer) was maintained throughout 1994 and 1995. At the central super-site, a detailed study of runoff processes, at scales ranging from small plots to catchment areas of several km^2 was conducted throughout the three-year monitoring period. The monitoring exercise also yielded a 1/200 000 scale map showing surface conditions, a knowledge of which is essential for testing hydrological processes in this area, as well as vegetation maps, land-use charts and maps of hydrological units at a smaller scale (1/50 000) but only for the super-sites. Before the start of HAPEX-Sahel, no such maps existed; the only documents available were a 1/200 000 scale topographical map and a 1/500 000 scale pedological map.

During the IOP, the ground meteorological parameters were measured over the one square degree. A set of 13 plots, distributed over the three super-sites, was equipped with instruments for flux measurements, and changes in vegetation (phenology, biomass, etc.) were monitored in detail. Detailed observations (of surface conditions, soil humidity, vegetation and aerosols) were made to validate the satellite data, coinciding in particular with aircraft flights and the passage of satellites with long return periods such as ERS-1.

All of the data collected in the course of HAPEX-Sahel, including the satellite data, have been integrated in a database that is freely accessible as of March 1997. The data are available on CD-ROM, and the database catalogue can be consulted on the World Wide Web (WWW). The server holds over 5000 structured and indexed data files.

Achievements of the experimental phase
During the HAPEX-Sahel programme, considerable progress has been made in a number of fields towards understanding hydrological and ecological regulation mechanisms. The results of this work will be published in a special issue of

the *Journal of Hydrology*, and are illustrated in Figures 6.3–6.7. However, due to the small size of the functional hydrological units, it was in the field of remote sensing that the most promising techniques were tested, contributing to a better overall understanding of the Sahelian climate. The use of microwave measurements was particularly interesting; it was demonstrated that the ERS-1 wind diffusion meter could be used to monitor the state of continental surfaces, and that vegetation and soil effects could be distinguished using a semi-empirical model.

Partial coverage of the three super-sites was obtained using the ARAT aircraft equipped with PORTOS, a multiple frequency, bipolar microwave radiometer operating at 5.1, 10.7, 23.8, 36.5 and 90 GHz with an adjustable beam angle of 0–50°. A ground measurement campaign was carried out at the three super-sites to accompany the airborne measurements, with soil moisture measurements on samples and by time domain reflectometry (TDR), and *in situ* measurements using the Marmotte radiometer.

All of the data were processed and navigated by high- and low-altitude flights and at all frequencies. It has been demonstrated that it is possible to measure soil moisture using a PBMR, a low-frequency radiometer operating at 1.4 GHz, and that PORTOS is sensitive to vegetation, even at the lowest frequency (5.1 GHz). In this case, the inversion of a simple model yields surface moisture values. Generally speaking, by altering the frequencies and the polarization and slope of the antenna, vegetation and soil effects can be distinguished. Work is now also in progress to assess the use of thermal infrared waves for estimating fluxes on sparse cover.

Microwave radiometry could play an important role in the spatial analysis of hydrological processes. In the more conventional area of visible and infrared wavelengths, interesting results have also been obtained from bidirectional reflectance measurements.

The development of vegetation was monitored using the POLDER radiometer (at visible and near-infrared wavelengths, with directional information, studies of aerosols with analysis by polarizing two wavelengths, and indications of water vapour with measurements at 930 nm) under highly variable optical thickness conditions, in the course of the ARAT flights over the three super-sites.

The ground campaign consisted of a number of sets of measurements: optical, the surface Bidirectional Reflectance Distribution Function (BRDF), and the radiation intercepted by vegetation. Significant progress is expected in determining vegetation parameters by remote sensing (fractional coverage, vegetation indices, etc.) from the POLDER data, and more generally, from the directional measurements. Some of the optical measurements (optical thickness, percentage of radiation intercepted by vegetation, etc.) are already being used for modelling. At the scale of one square degree, the BRDF estimates are now being used to calculate the albedo.

Appraisal and outlook

The contribution of the EPSAT-Niger and the HAPEX-Sahel experiments to our knowledge of the processes that control the interactions between climate and hydrology in dry tropical areas has so far been restricted to the local level. However, a number of larger-scale questions, such as the dependence of aquifer recharge on the abundance of rainfall and the timing of the rainy season, are also being addressed. In particular, it is apparent that the recharging of the upper aquifer – the main water resource for the Sahel – depends only in part on the amount of rainfall during the rainy season. The upper aquifer is recharged mainly by temporary ponds. Since these ponds are filled by surface runoff, the annual recharging of the aquifer requires rainfall conditions that favour runoff, i.e. rainfall concentrated in short periods of time. As well as successfully reproducing rainfall totals, it is therefore essential that the GCMs also simulate realistically the distribution of rainfall over the rainy season. In this respect, the data from the EPSAT-Niger experiment provide an important means of validating the realism of the models, at least for the present period. The processing of the HAPEX-Sahel data has only just begun, so that there are likely to be suggestions for more accurate parameters (that offer a better reproduction of the distribution of evaporation, runoff and infiltration), in order to describe the processes by which water is redistributed in the soil.

In addition to these achievements, which have already been proven or are currently under development, major difficulties have arisen in some areas. Thus it may be necessary to schedule additional work in order to continue, or to add to by adopting different approaches, the work carried out in Niger in 1990–93. The most serious difficulties appear to be related to specific features of the environment in semi-arid areas and their impacts on meso-scale 3D models. For example, sudden transitions caused by passing convective systems can result in inaccurate specifications of surface moisture. On the other hand, it is impossible to validate properly the hydrological balance calculated by an atmospheric model due to the lack of functional hydrological integrators at scales corresponding to the resolution of the models (from 10×10 km for meso-scale models, to 200×200 km for the GCMs).

Another difficulty is related to the instruments, although this could, to a large extent, be remedied by appropriate

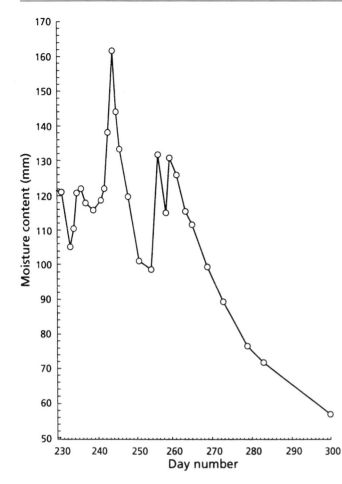

Figure 6.3 Soil moisture depletion during the IOP in the top 1.5 m of a millet field at the southern super-site (data provided by the Institute of Hydrology, Wallingford, UK).

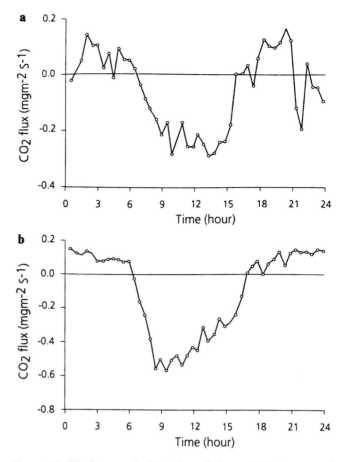

Figure 6.5 CO₂ fluxes at the West Central site on (a) 21 August and (b) 9 October (data provided by the Winand Staring Centre, Wageningen, The Netherlands).

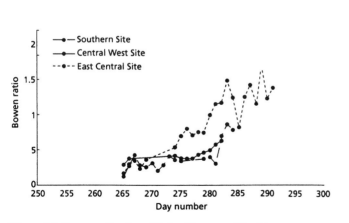

Figure 6.4 Bowen ratios (sensible heat/energy utilized by evaporation) during the IOP at the three super-sites after the last rains (data from the Institute of Hydrology; ORSTOM; Winand Staring Centre).

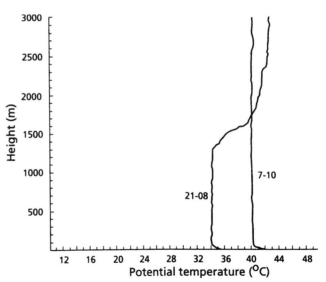

Figure 6.6 Radio-sounding profiles at the Hamdallay site at the start (21 August 1992) and end (7 October 1992) of the IOP (data provided by CNRM, Toulouse).

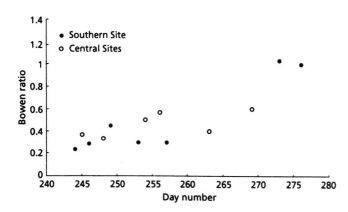

Figure 6.7 Bowen ratios obtained from the Merlin flux aircraft for the Central and Southern super-sites (data provided by METEO-France/CNRM/CAM and Laboratoire d'Aerologie, Toulouse).

effort. There is a shortage of atmospheric data to serve as a framework, due to the very small number of operational radio-sounding stations in the Sudano-Sahelian strip. The lack of observations prevents the quantification of advection at the HAPEX-Sahel site, even though due to the lack of relief and surface water reservoirs, this represents a major source of atmospheric forcing. However, the intensive radio-sounding operations carried out during the IOP have demonstrated the good fit between vertical fluxes, at the local level, and the soundings. The relative simplicity of atmospheric circulation in the Sahelian region could therefore be monitored by doubling or tripling the number of permanent radio-sounding sites. The resulting density would nevertheless remain significantly lower than in temperate areas. Although it is perhaps not relevant to the subject of this chapter, the inadequacy of the radio-sounding network represents a serious handicap for analysis and realistic forecasting in tropical areas using meteorological models, which are no more than operational versions of the GCMs.

Other experiments of the same type as HAPEX will be necessary to improve our understanding of the surface–atmosphere interactions in other African climate systems, such as the Sudan, the coastal areas, and areas of central Africa where the relief plays a major role. In the coming years, the increasing amount and diversity of remote sensing data will undoubtedly lead to the reappraisal of numerous aspects of the experimental programmes decided for these projects. Nevertheless, their overall philosophy should be conserved, if only on account of the large differences in the scale of the fluctuations related to hydrological and atmospheric processes. The joint study of these fluctuations would require more intensive use of instruments, which can cover

only limited areas for restricted periods of time. Within the framework of GEWEX, there are plans to set up a project that would combine an observation area of several tens of thousands of km^2, affected by the Sudanese climate, with intensive experiments using a protocol similar to that used for HAPEX-Sahel. Potential sites for the project are currently being examined, and two areas have been singled out: the Oueme basin in Benin, and the Bandama or Comué basins in the north of Ivory Coast. In addition to the fact that the long-term monitoring period will last for at least five years, there is one significant difference compared with HAPEX experiments in the past – a catchment basin has been chosen as the experimental contour. From the point of view of hydrological integration, this is more consistent than working on a square (the choice of one square degree as the basis for the first HAPEX experiments was the result of atmospheric modelling constraints, since GCMs use square grids). With a basin covering several GCM squares and extending in a north–south direction, it will be possible to explore, on the one hand, inter-mesh variability and, on the other, the influence of the climatic gradient over much greater distances than was the case in HAPEX-Sahel. Furthermore, by operating in the Sudanese climate, there would be the guarantee of perennial runoff in basins covering more than 1000 km^2, which would permit genuine hydrological validation. Finally, the rainfall systems for these areas are more varied than those observed in the Sahel, which will be an asset for providing ground-based references for the TRMM and Tropiques satellite missions.

6.3 EAST AND NORTHEAST AFRICA

This section describes the macro-scale model results obtained for East Africa, the Nile basin, and the Zambesi basin.

6.3.1 Macro-scale model results for East Africa

In a preliminary investigation of the effects of global warming on hydrological regimes in East Africa, Arnell and Reynard (1995) used a gridded macro-scale hydrological model applied at a resolution of $0.5 \times 0.5°$. Each grid cell, covering an area of approximately 3000 km^2, was treated as an independent catchment, and a daily conceptual water balance model was applied in each cell. The model, derived from that of Moore (1985), assumes that the field capacity varies across the catchment (or grid cell) following a statistical distribution. Model parameters include an average catchment field capacity (determined from digitized soil texture maps), and a parameter defining the form of the statistical distribution (assumed to be constant). In calculating interception losses, the model distinguishes between forest and grass cover, and includes two routing parameters. The model operates on a daily basis, using daily precipitation generated statistically from the gridded monthly precipitation, and daily potential evaporation: the resolution of $0.5 \times 0.5°$ was dictated by the available resolution of the input data on precipitation and potential evaporation.

Model validation showed that the model simulated the annual water balance reasonably well across much of East Africa, as far as could be determined from the available data, and that major discrepancies could be explained in terms of the presence of swamps and inadequacies in the input precipitation data.

Two types of studies were undertaken with the model. The first was a sensitivity analysis, which examined the effects of changes in precipitation of -20, -10, $+10$ and $+20\%$, and increases in potential evaporation of 5, 10, 15 and 20%. Figure 6.8 shows the sensitivity of annual runoff to different combinations of changes in rainfall and potential evaporation, for four representative cells. The most general conclusion is that runoff is much more sensitive to changes in rainfall than to changes in potential evaporation (the contours on the sensitivity plots are nearly vertical); this is because evaporation rates in the region are determined by water availability, and not by potential demand. The spacing between the contours reflects the sensitivity to changes in rainfall. In general, the lower the runoff coefficient, the greater is the sensitivity to change (as has been shown in many other studies). With a runoff coefficient of 0.1, and in the absence of changes in potential evaporation, a 20% increase in rainfall would result in an increase in annual runoff of 100%; with a runoff coefficient of 0.4, the increase in runoff would 'only' be around 40%. This general relationship breaks down in very arid catchments, where very little of the rainfall actually becomes runoff. The relationship between the changes in rainfall and in runoff is generally linear (the contours are straight), although in the driest cells there is a tendency for an increase in rainfall to have a larger relative effect on runoff than a decrease in rainfall.

The second set of studies was based on climate change scenarios developed by four global circulation models. The pattern of change in runoff is highly dependent on the pattern of change in rainfall. Runoff is simulated to increase over large parts of Uganda, Kenya, eastern Tanzania and northeast Mozambique, but to decrease further west. Different climate models produce different patterns of change in runoff.

These studies, although preliminary, illustrate two main points. First, hydrological regimes and water resources in East Africa are sensitive to changes in climate, and particularly to changes in rainfall; an increase in temperature alone has relatively little effect across the region. Second, there are considerable differences between climate change scenarios, largely reflecting the differences in simulated changes in rainfall. Until rainfall scenarios are refined, the estimates of the effects of climate change on water resources in East Africa will remain highly uncertain.

6.3.2 The Nile basin

The Nile basin drains approximately 10% of the African continent (2.9 million km^2), is shared by nine countries and provides freshwater for millions of people, at least half of whom live in arid zones without local surface water resources. The freshwater resources originating in the upstream highland countries of the basin are vitally important for all the countries of northeast Africa, but particularly for Egypt, which is highly vulnerable to future changes in water availability. This hyper-arid country already takes its full water allocation as defined by the 1959 Nile Waters Agreement, and in some years a little more. This allocation (66% of the average natural flow of the Nile at Aswan) corresponds to $55.5 \, km^3$ per year, an amount based on water yields in the period 1900–59. Balancing the supply of scarce water, which could be affected by climate change, with increasing water demand, under the pressure of growing population and development, will clearly be a tremendous concern in the coming decades. The Nile basin is also one of the most complex major river basins of the world, due to its size, the variety of climates, the topography of the sub-basins, and human influence (such as dams and reservoirs). This complexity, and the paucity of available data, illustrates the difficulties involved in developing reliable sophisticated models taking into account possible future climatic conditions.

Paleoclimatic background

It is thought that the White and Blue Nile basins did not merge until the early Pleistocene and that the summer floods were not established before 25 000 BP (Evans, 1994). During the last glacial maximum (*Würm*) the area of the upper Nile received much less rainfall than it does today, and the basins of Lakes Victoria and Albert were closed before 12 500 BP. The Sudd region was very dry and the White Nile was blocked by dunes south of Khartoum (Saïd, 1994). After

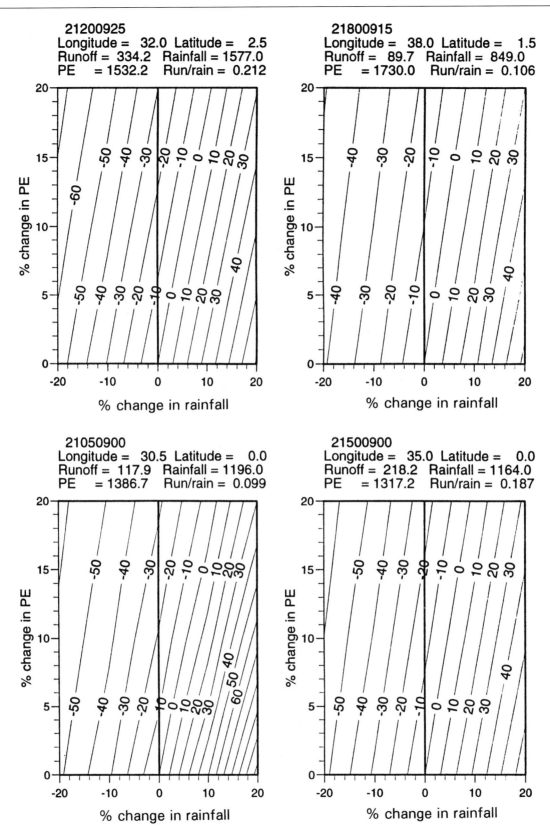

Figure 6.8 Sensitivity of annual runoff to changes in precipitation and potential evaporation (mm/yr), for four selected grid cells in East Africa (Arnell and Reynard, 1995).

12 500 BP, there was a period of increased rainfall, and Lakes Victoria and Albert overflowed into the Nile. According to Street and Grove (1979), the period 10 000–8000 BP was very wet with high lake levels at the end of the period. After 7500 BP, numerous trend reversals occurred, but with a progressive desiccation leading to semi-desert conditions.

Dynastic Nile records and the Roda Nilometer

Despite its mysteries – the location of its sources, the predictability of the rise and fall of the annual flood – the case of the Nile is unique in that extensive information on river levels has been recorded on cliffs, stones and the pillars of temples since as far back as 3000 BC. However, this information is difficult to analyse, and consistent reconstructions of flood levels remain impracticable. Egyptian civilization under the pharaohs was closely dependent on the amplitude of the annual flood, and the short-term sequences of droughts or floods had a profound influence on the fortunes of the different dynasties. According to Bell (1971), the first 'Dark Age', which began around 2150 BC, was the result of short but intense sequences of droughts; the period 1990–1570 BC was one of high floods; and the second 'Dark Age' (1200 BC) was due to severe drought conditions. Egypt then entered a prolonged period of decline with increased scarcity of water.

The most famous record of Nile flood levels is the Roda Nilometer, which was found on Roda Island near Cairo, in the middle of a well connected with the Nile by conduits. The lines engraved on the shaft of this octagonal white marble column represent a record of Nile levels since 641 AD. Numerous studies have been made of these records (e.g. Jarvis, 1936) and they have been used to correct various inhomogeneities (zero change, siltation processes, etc.). The most exhaustive study was undertaken by Popper (1951), and Hurst (1951) used these records to elaborate his K coefficient (assuming that the sequences of low and high flows tend to occur together). From the tentative efforts that have been made to convert these recorded levels to discharges, it appears that since AD 1720 the Nile flows have become more variable, and extreme flow sequences have become much more acute and persistent (Evans, 1994).

Instrumental period

Rainfall

As for the other main river basins of Africa, the Nile basin has experienced considerable rainfall fluctuations since the beginning of this century and extensive periods of drought, such as those of 1913 and 1983–84. Due to its latitudinal extent, the rainfall regimes affecting the basin vary from desert to equatorial. According to Conway and Hulme (1993), for a comprehensive study of rainfall–runoff relations, two broad homogeneous regions must be considered: the Ethiopian highlands (the Blue Nile and Atbara), and Lake Victoria and the equatorial lakes (White Nile), which have had contrasting precipitation regimes uncorrelated in time. In a study of African rainfall changes between 1931–60 and 1961–90, Hulme (1992) described the different behaviours of the Sahel region, with a continuous rainfall decline in recent decades, and the equatorial East African region where after the extremely wet years of the early 1960s (which caused an abrupt rise in the level of Lake Victoria), the rainfall anomalies remained slightly positive. After calculating the average annual rainfall over 40 years (1945–84) in the main sub-basins of the Nile, Conway and Hulme (1993) compared two successive sub-periods of 20 years. Table 6.3 shows the rainfall evolution in these regions (obviously, these estimates are questionable due to the low station density or poor data accuracy).

Runoff

The mean annual flows of the Nile at Aswan (naturalized, i.e. corrected to account for human abstraction, reservoir losses, etc.) have decreased considerably over the last century. In the period 1871–89 the mean annual flow was 88 km^3. The 1959 Nile Waters Agreement was based on an annual flow of 84.5 km^3, but by 1972–89 (the period affected by the Sahelian drought) it had already fallen to 77.2 km^3 (i.e. a deficit of 9%), and by 1983–87 to 68.4 km^3 (a 20% deficit). The weakest annual flow observed since 1913 was in 1984, as in many major rivers, with only 42 km^3 (i.e. 58.2 km^3 naturalized). Lake Nasser, created by the High Aswan dam, has a storage capacity of 134 km^3; it started to fill in 1963, and was almost full by 1978. Since then, however, lake levels have fallen dramatically and in 1988, before the heavy rains at Khartoum, the reservoir was near its dead storage level (i.e. storage volume that can not be released).

However, a single analysis of the main Nile discharges can not explain satisfactorily the response of Nile runoff to rainfall variations over past decades. The rainfall regimes over the Blue and White Nile head catchments are different and they have evolved in different ways during the Sahelian drought. Conway and Hulme (1993) calculated the percentage changes in runoff depth from a comparison of rainfall in the periods 1965–84 and 1945–64 (see Table 6.3). In the most recent period (1965–84) the sub-basins of Lake Victoria and the equatorial lakes have shown increases of 60 and 59%, respectively, whereas runoff in the Blue Nile and Atbara sub-

Table 6.3. *Annual rainfall in the main river basins of East Africa, 1945–84 (Conway and Hulme, 1993).*

Sub-basin	Area (km²)	Average annual rainfall (mm/yr)	% Difference, 1965/84–1945/64
Lake Victoria	258 000	1356	+ 2
Equatorial Lakes	293 000	1164	− 7
Sudd	129 000	935	− 10
Bahr el Ghazal	475 000	859	− 11
Sobat	231 000	1191	− 14
Central Sudan	310 000	518	− 11
Blue Nile	195 000	1224	− 8
Atbara	137 000	515	− 11

basins have decreased by 18 and 32%. According to Sutcliffe and Lazenby (1994), in 1965–84 the annual flow of the Blue Nile at Khartoum was 45.6 km³ (− 15% compared with 1912–64), whereas that of the White Nile at Mogren was 33.9 km³ (+ 32%). The decline of the Blue Nile flow has therefore been partially alleviated by the higher floods of the White Nile.

Difficulties in assessing climatic change in the Nile basin

The High Aswan dam with its over-year storage has allowed the regulation of the annual main Nile inflows and the maintenance of the supply of water to Egypt in spite of the intense drought. However, it is not possible to determine whether the recent sharp variations in active storage indicate the effects of short-term climate variability or the prospect of more persistent longer-term change. The difficulties involved in studies of the potential impacts of climate change have been highlighted in Section 6.2 in relation to West and Central Africa, but the Nile basin presents a special challenge in view of two compounding factors. First, from the hydrological point of view, there are several major lakes in the highlands, large areas of permanent or seasonal wetlands, and large reservoirs; and second, from the climatic point of view, with the wide range of climates, ranging from mountainous equatorial conditions with abundant precipitation, to northern arid plains with rare precipitation. These two contrasting rainfall regimes, due to quite distinct meteorological conditions, explain the differences in the behaviour of the two branches of the Nile. The Blue Nile is highly sensitive to a monsoonal rainfall regime, whereas the White Nile is fed by equatorial rains with a bimodal seasonal distribution. The lack of correlation in either rainfall or runoff between the two branches of the Nile is of great importance in modelling the basin as a whole.

As noted by Conway (1994), there have been numerous studies to assess the impacts of fluctuations in Nile discharges on water resources, but few have attempted to evaluate the impacts of future climate change on runoff. In a study of rainfall–runoff relations during the instrumental period, Conway and Hulme (1993) divided the basin into eight sub-basins and demonstrated the importance of the Blue Nile and Lake Victoria sub-basins. Gleick (1991) showed that is necessary to elaborate and calibrate hydrological models (using, for example, the theoretical water balance model elaborated by Wigley and Jones, 1985) with the Nile basin carefully split into sub-basins representing the different physical, hydrological and climatic characteristics of the whole basin.

In the following we describe some attempts that have been made to calibrate hydrological models using the surface air temperature and precipitation produced by GCMs for simulating the effects of a doubling of atmospheric CO_2 on the Nile basin.

Climate change modelling results

Hulme (1990) described the various possible approaches for evaluating future precipitation over the Nile basin: short-term seasonal forecasts using SST anomalies (although these may not be relevant for the Ethiopian or Ugandan catchments), analyses of instrumental time series data, historical analogues, and the use of climate scenarios from GCM experiments. Using this last approach, Santer *et al.* (1990) developed a composite GCM scenario derived from five models (GISS, GFDL, Oregon State University (OSU), NCAR and UKMO), assuming a doubling of atmospheric CO_2. Computed maps for different parts of the world show the evolution of two variables (surface air temperature and precipitation) for two seasons (December to February, DJF, and June to August, JJA). Hulme extracted the following results for East Africa. The air temperature can be expected to increase by about 3–4°C in DJF, and by about 2–3°C in JJA. For precipitation, there is no evidence of change for Ethiopia or central Sudan, but for Uganda there is a high probability of an increase (36 mm or 28% in DJF, and 63 mm or 18% in JJA). These contrasting results for the two main Nile catchments are in accordance with the present meteorology.

To test the vulnerability of runoff to changes in precipitation, Gleick (1991) combined meteorological information on future climatic changes with the use of a regional hydrological model. In a first step, he developed a simple annual water balance model by splitting the Nile basin into six sub-basins. Three of these (the Sudd swamp, the White Nile between the Sobat river and Khartoum, and the lower Nile from Aswan to Delta) generate very little runoff, but the other three (the Upper White Nile, the Sobat River, and the Blue Nile/Atbara regions) produce substantial runoff. In a second step, using the estimates of changes in temperature and precipitation given by the GFDL model, Gleick (1991) found a 30% increase in runoff over the Upper White Nile region due to increased precipitation in the basin, whereas in the Blue Nile/Atbara he found a 50% decrease in runoff due to a 20% decrease in precipitation. For the whole of the Nile basin, he found a 25% decrease in runoff.

Conway (1993) developed a grid-based hydrological model on a monthly time scale to evaluate changes in discharges over the Blue Nile basin ($170\,000\,\text{km}^2$). The model was calibrated to reproduce the actual mean monthly runoff, and the simulated mean annual runoff was within 3% of the observed value. In three GCM experiments (GFDL, GISS, Composite), based on an increase in mean global temperature of 1°C, it was assumed that this would give a 4% increase in potential evapotranspiration (PET, based on Thornthwaite). The changes in annual precipitation ranged from -2% (GFDL) to $+9\%$ (Composite), and the changes in annual runoff were -8.6%, $+15.3\%$ and $+0.7\%$, respectively. In a similar study of the Lake Victoria basin, using a water balance model developed by the Institute of Hydrology (UK), Gleick (1991) found that the changes in PET were more important (due to the expanse of open water), and the changes in runoff were -9.2%, $+11.8\%$ and $+6.9\%$, respectively.

In the research project 'Complex River Basin Management' sponsored by the US EPA (Saleh *et al.*, 1994), a hydrological model was constructed on a monthly time scale, and every sub-catchment was modelled in a lumped, one-dimensional water balance model with spatially weighted parameters. The climate changes were simulated by the GFDL, GISS and UKMO GCM scenarios. With a doubling of CO_2, the increases in air temperature were 3.2%, 3.4% and 4.7%, and the changes in runoff at Aswan were -77%, $+30\%$ and -12%, respectively. Following EPA guidelines, the sensitivity of the model was tested in scenarios with arbitrary temperature increases of 2° and 4°C, with no change in precipitation and $\pm20\%$ changes in precipitation. The results showed considerable variations; for example, Nile flows decreased by 98%(!) with a 4°C warming and a 20% decrease in precipitation, and with no change in temperature the runoff varied between -73% and $+71\%$ with $\pm20\%$ changes in precipitation.

As noted by Conway (1994), it is not yet possible to produce reliable estimates of future Nile flows taking into account the effects of climate change. However, the studies undertaken in the last decade have improved our understanding of the behaviour of the Blue Nile and Lake Victoria sub-basins. The sensitivity analyses indicate that the Nile basin is extremely sensitive to any change in climate, but the results obtained so far may not be realistic because the models do not incorporate changes in vegetation or soil properties.

6.3.3 The Zambezi basin

The Zambezi basin was also selected as part of the research project 'Complex River Basin Management', to evaluate the impacts of climate change following EPA guidelines. The results are to be published by Cambridge University Press.

The Zambezi River basin is the fourth largest in Africa and is shared by eight countries. The Kariba dam on the main river is one of the largest reservoir systems constructed in recent years, with an active storage capacity of $70\,\text{km}^3$, and is a vitally important source of energy for Zambia and Zimbabwe (Campos, 1989). For this reason, the problem of the sensitivity of the water storage system to climate change is critical (Kaczmarek, 1990). A set of scenarios has been developed to determine direct changes in runoff parameters on the basis of past hydrological and meteorological observations. It is expected that, with a doubling of CO_2, the average inputs to the Kariba reservoir will vary by no more than $\pm20\%$ from those of the past 60 years (Salewicz, 1995).

6.4 CONCLUSIONS AND RECOMMENDATIONS

Africa's hydrological regimes vary widely from region to region, and the continent's water resources are very unevenly distributed. This lack of uniformity may lead to disputes between countries in the near future unless a balanced sharing of these resources can be planned.

The continent's climate has fluctuated widely since the last glacial maximum, and over the past several decades a number of African countries have suffered from the persistent drought that has led to extensive desertification. The accumulated effects of such a water deficit on the hydrological regimes and on the environment can provide some very valuable empirical lessons. For political leaders and decision-makers, these effects also demonstrate the fragility of the large-scale water engineering structures that were built in the 1960s on the basis of more favourable yield data series.

There is a wide gap between the results from hydrological models and those from GCMs, due to the inadequate knowledge of soil/vegetation/atmosphere exchanges at the largest scale. This explains why current studies produce such divergent results with regard to runoff. The HAPEX–Sahel experiment, which was designed precisely to fill this gap, is the first large-scale operation of this kind and is particularly promising. But it also shows that similar progress in other climatic zones will depend on other experiments of the same kind being performed.

These conclusions lead to the following recommendations:

- In instrumental terms, reliable observation networks must be maintained (for hydropluviometry in particular), in order to acquire high-quality, homogeneous time series reference data. Given the scarcity of available resources and the need for regular monitoring, it is essential to foster operations like the World Hydrological Cycle Observing System (WHYCOS), which can compensate for possible local deficiencies.

- In historical and paleoclimatic terms, a detailed knowledge of the climate record of the past is essential for assessing potential climatic fluctuations. Improved dating methods should provide more accurate information on the pace of climate change since the last glacial period (particularly abrupt changes). The IGBP's PAGES (Past Global Change) project and the European Union's PEP3 (Pole–Equator–Pole) project for Europe

and Africa should therefore be encouraged. PEP3 will analyse and correlate the paleoclimatic records of both hemispheres as part of the PANASH (Paleoclimates of the Northern and Southern Hemispheres) programme. Similarly, the studies of forest ecosystems being conducted under the IGBP's Global Change and Terrestrial Ecosystems (GCTE) programme should make a useful contribution to our understanding of ecosystem changes in the equatorial forests.

- With regard to the models, so far there have been too few case studies, in either the field of hydrological modelling or GCMs. To improve the GCM results, we need a much more thorough knowledge of spatial and temporal variations in water (in the atmosphere and on and in the ground), as well as more accurate estimates of the parameters involved in the relations between the soil, vegetation and the atmosphere. It will therefore be necessary to conduct further HAPEX–Sahel-type projects in other climatic zones of Africa (such as in coastal, humid and mountainous regions).

- For decision-makers, policies on water resource allocation and official standards for major water engineering structures must be decided prudently, with emphasis on consultation among all water users and scientists (the recent increases in the runoff coefficients of small basins in the Sahel, despite the overall rainfall decline, is a good example of the need for consultation: this unexpected phenomenon is linked with the evolution of soil properties and disappears when the basin size exceeds $20\,000\,km^2$ due to the effects of evaporation and hydrographic degeneration).

- For scientists, it is essential that more African scientists are involved, and to a greater extent, in studies of climatic variability, climate trends and their impact on the hydrological systems of Africa. In this respect, an important step is being taken in the FRIEND projects (Flow Regimes from International Experimental and Network Data), which are now operating in Southern Africa, West Africa, and the Mediterranean region (and the launching of the Nile project).

7 Assessment of the impacts of climate variability and change on the hydrology of Asia and Australia

I. A. SHIKLOMANOV and A. I. SHIKLOMANOV

7.1 INTRODUCTION

The continent of Asia covers an area of 43.5 million km^2, one-third of the Earth's land surface, and is home to more than half of the world's population. The vast extent of the continent – more than 10 000 km west to east – with its high mountain ranges and extensive lowlands, together with irregularities in atmospheric circulation patterns, cause uneven distributions of heat and moisture. There are permafrost zones, large river systems with areas of water surplus, and hot dry deserts.

Over most of Asia, particularly in the west and south, the distribution of precipitation is uneven, and in some areas the annual water deficit attains 1500–2000 mm. Asia is the world's major user of freshwater, and water resources are generally well developed; more than two-thirds of the world's irrigated land is located here. Total surface water resources amount to 14 400 km^3/year, and water availability per capita is 13.1 m^3/day, or half of the world average. Shiklomanov and Markova (1987) estimated that in 1990 water consumption for economic uses in Asia would be 2400 km^3, or 59% of the world total. This represents 17% of annual streamflow, an amount that could increase to 20–25% in the near future. The development of agriculture, the main source of economic prosperity, will be limited by the availability of water. Reliable assessments of possible changes in climate due to global warming, and the effects of these changes on hydrological regimes and water resources, are therefore vital for economic planning in many Asian countries.

The influence of climate on water resources is also important for Australia, the smallest continent by area and by population. The continent is extremely dry but its agriculture is highly developed. River water resources are limited (300 km^3/year), but due to the small population the per capita water availability is high, about 50 m^3/day, or twice the world average. Meanwhile, the mean specific discharge is very low, i.e. 1.4 l/s km^2. Except for a narrow coastal area in the northeast, the continent receives insufficient precipitation, and there is a high natural water deficit of up to 2000 mm/year in the interior; deserts and semi-deserts cover about 75% of the continent. In most regions agriculture is dependent on irrigation; most of the 1.9 million hectares of irrigated farmland are located in the Murray River basin, the only significant river system. Any further development of irrigation will be restricted by the availability of water resources.

Under the conditions of water deficit that characterize many parts of Asia and Australia, hydrological systems and water resources are sensitive even to minor changes in climate and human activity (Shiklomanov and Lins, 1991). Thus the interaction of climate and hydrological characteristics, particularly in assessments of the hydrological impacts of global warming, is very important for these regions.

This chapter deals with three aspects of climate change in relation to Asia and Australia. Section 7.2 examines the interaction between climate and hydrological characteristics in the Caspian Sea basin, on the basis of historical data and long-term observations. Possible changes in hydrological characteristics due to climate change are also assessed for other regions of Asia, particularly in light of the observed increases in global air temperatures in the 1980s. Section 7.3 analyses the published estimates of the effects of climate change on hydrological characteristics and water resources in the region. Section 7.4 describes GEWEX-GAME, a long-term international study of water and energy cycles in the monsoon regions of Asia and Australia. This experiment aims to provide insight into the effects of climate change on components of the hydrological cycle in the region. Finally, Section 7.5 presents some conclusions and recommendations for future research.

85

7.2 INTERACTIONS BETWEEN CLIMATE AND HYDROLOGICAL CHARACTERISTICS

7.2.1 The Caspian Sea basin

The Caspian Sea is the world's largest closed water body. At present the water surface is about 27 m below mean sea level, with an area of 376 500 km^2 and a volume of 78 290 km^3. The sea receives waters from the rivers of Azerbaidjan, Iran, Kazakhstan, Russia and Turkmenistan, which drain an area of about 3 million km^2. These rivers include the Volga and Ural (1.6 million km^2); the Kuma, Terek, Sulak and Samur rivers of the Caucasus and Trans-Caucasus (254 000 km^2); the rivers of northern Iran (103 000 km^2); as well as smaller rivers between the Volga and the Ural, and from the Ural to the Iranian border (819 000 km^2; see Figure 7.1).

The level of the Caspian Sea fluctuates considerably due to the variability of natural hydroclimatic factors in the basin and the water area, which depends on the inflowing river runoff, precipitation, and evaporation from the water surface. The level of the sea is an integral characteristic of the total volume of water in the basin and surrounding areas. Changes in air temperature in these regions inevitably lead to changes in sea level. The Caspian Sea basin provides an excellent example of the close relations between variations in climate and hydrological characteristics that can be observed from paleoclimatic and historical records, as well as from recent instrumental data.

Paleoclimatic and historical data

Recent paleoclimatic and paleohydrological studies of the Caspian Sea basin (Klige, 1985; Varushchenko *et al.*, 1987) have been based on geological and archeological data, analyses of lake deposits, and studies of the oxygen isotope compositions of ice cores. With these data, a detailed picture of the patterns of change in the hydrological cycle in the northern hemisphere has been built up, particularly in relation to the level of the Caspian Sea since the last so-called Valday glaciation in Europe.

During the last Ice Age, about 50 000 years ago, air temperatures in Europe were 5–10°C lower than today. The Caspian Sea was more than 75 m above its present level, with an area of 800 000 km^2 and a volume of 130 000 km^3. According to Kalinin *et al.* (1966), annual river runoff into the Caspian Sea amounted to 545 km^3, approximately twice as much as today, and annual evaporation was 680 mm, about 40% less than today. During the last interglacial,

23 000–25 000 years BP, the climate was warmer and drier, and all closed lakes in the northern hemisphere receded. At this time the Caspian Sea level was probably the same as or lower than it is today.

The variations in the level of the Caspian Sea and the determining climatic factors throughout historical time (within the last 2500 years) have been cyclical in nature, due to changes in water inflows and net evaporation, as shown in Figure 7.2 (data from Klige, 1994). In the last 2200 years there have been seven cycles of elevation and lowering of the sea level, each lasting 300–400 years. It is believed that the lowest sea level (−30 to −33 m) occurred in the sixth century, and the highest (−23 to −24 m) in the late eighteenth/early nineteenth centuries, giving a maximum fluctuation amplitude due to natural climate variations of about 8–10 m. The lowest level corresponds to reduced inflows and increased evaporation, and vice versa. According to Zaikov (1946) the amplitude of sea level variations over the last 450 years (since 1550) due to climatic factors has been 5.5 m.

In view of the cyclical nature of secular variations in the Caspian Sea level (Figure 7.2), it appears that we are now at the end of a period of regression that has lasted for almost 200 years. In the future, there is likely to be a slow elevation of the sea level, and this has indeed been observed during the last 15 years.

Instrumental observations

Instrumental observations of the Caspian Sea level began in 1830 at Baku, and regular measurement have been made since 1837. At the turn of the century gauge stations were established at many sites along the coast, and meteorological and hydrological stations elsewhere in the basin. From these data reliable estimates can be made of variations in the terms of the water balance of the Caspian Sea.

Figure 7.3 shows the variations in mean annual sea level between 1880 and 1995. In the first 50 years of this period the sea level was generally high, with a slight lowering trend. The maximum level was recorded in 1883 (−25.3 m), and the minimum in 1925 (−26.6 m), giving a maximum fluctuation of 1.3 m. Between 1933 and 1940 the sea level fell dramatically by 1.7 m, followed by continued but less rapid lowering. In 1977 the level was −29.10 m, the lowest since instrumental observations began. Since 1978 the sea level has gradually risen, reaching −26.65 m in 1995, an increase of 2.45 m in just 17 years. Since the late 1950s the Caspian Sea has been affected by human activities (e.g. water abstraction for irrigation, industrial and domestic consumption, and the

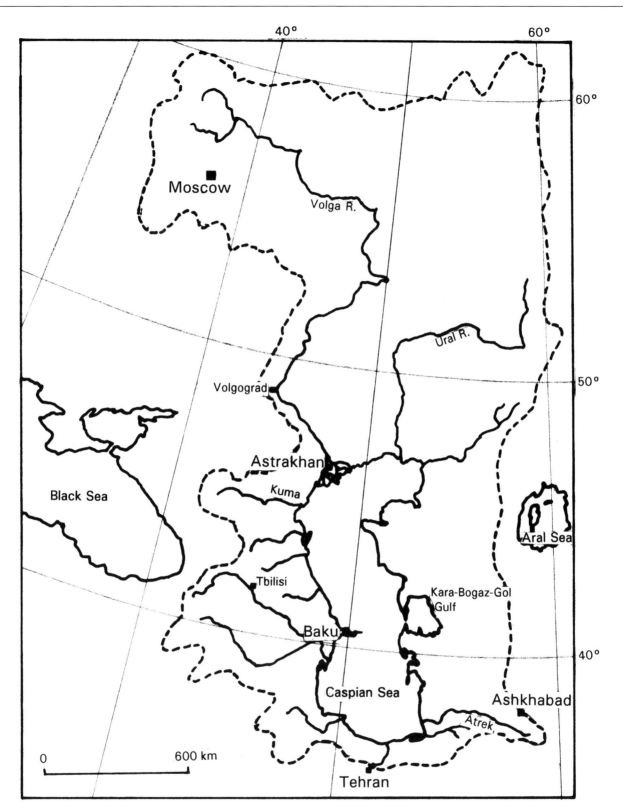

Figure 7.1 The Caspian Sea basin.

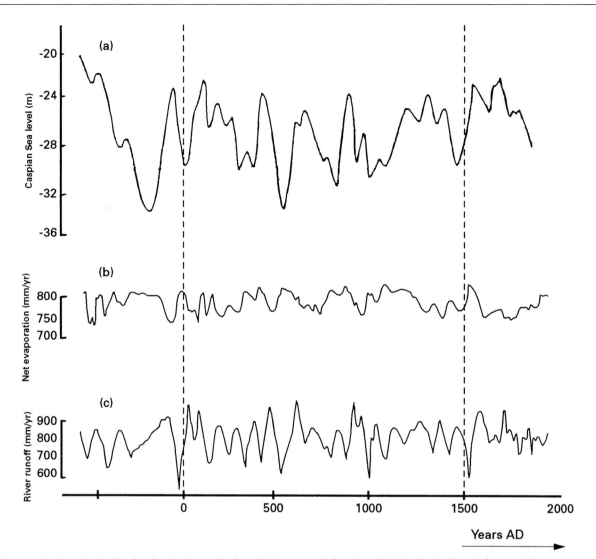

Figure 7.2 (a) Variations in the Caspian Sea level over the last 2500 years, and the determinining hydroclimatic factors; (b) net evaporation, and (c) river runoff (from Klige, 1994).

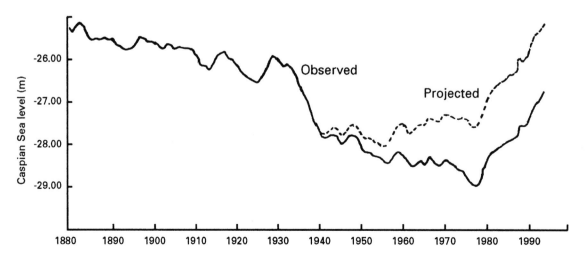

Figure 7.3 Observed Caspian Sea levels, 1880–1995, and projected level in the absence of human interference (dashed line).

construction of dams on the rivers feeding the sea, both of which have reduced inflows), and increased water losses due to evaporation from the sea surface. Without this human interference, the sea level would have been considerably higher than it is today (dashed line in Figure 7.3).

The long-term variations in the Caspian Sea level are determined by the relations between climatic factors in the basin, and most recently by human activities. Quantitatively, the role of climatic factors can be seen from analyses of the components of the water balance over the long term. Table 7.1 presents details of the Caspian Sea water balance since 1880, based on estimates by the Russian State Hydrological Institute (SHI).

Between 1880 and 1913 the sea level fell by 92 cm, even though the total inflows were above the mean (306 km^3/year), and the average visible evaporation (evaporation minus precipitation) was below the mean (71.5 cm/year). This is explained by the fact that to maintain the average high water level (-25.5 m) of the 1880s, even with visible evaporation below normal, the total inflows to the sea should have been about 312 km^3/year. Between 1914 and 1932 the sea level rose slightly, and between 1933 and 1940 it fell dramatically by 172 cm, the largest fall since observations began. This can be attributed to reductions in incoming terms of the water balance, especially inflows, which were about 78% of normal, and to increased evaporation. In subsequent years (including 1974) the sea level continued to fall due to reduced inflows resulting from human activities, such as water abstraction for irrigation, for reservoirs, and increased evaporation from the water surface (Shiklomanov, 1976). The most significant fall occurred between 1971 and 1977, when the average annual inflows were only 236 km^3. In this case, human activities in the basin led to a reduction in annual inflows of 28 km^3 (Shiklomanov, 1988).

Up to the late 1970s, the gradual fall in sea level observed over the previous 100 years had given rise to the illusion that this was an ongoing process that would continue in the future. The coastal areas were developed intensively, with residential and recreational areas, industrial units and agricultural enterprises. In the 1970s the most acute water management problem in the former Soviet Union was to prevent further reductions in the Caspian Sea level and to stabilize it. Grandiose projects to divert parts of northern and Siberian river discharges to the Volga basin were widely discussed. In 1978, however, the sea level began to rise again, and by 1994 had risen by 229 cm, accompanied by widespread flooding of large areas around the sea, and increased pollution of water supplies. The situation was critical. Many buildings and

structures in the coastal region have been ruined, and roads, beaches and valuable agricultural lands have been inundated. The higher sea level has led to increased erosion by wind and storm waves, and has also caused changes in hydrometeorological and ecological processes in the coastal region, particularly in delta regions.

Urgent measures are now needed to protect the coastal areas and to rectify the damage to natural ecosystems, and to implement an efficient environmental monitoring system. To design and develop a general strategy to ensure an effective water management system under the conditions of variations in sea level, reliable estimates are needed of the potential changes in the future. To elaborate such forecasts, however, it is necessary to study in detail the dynamics of the Caspian Sea water balance in the past, and to determine the causes of the recent sea level rise.

Causes of sea level rise

There is little consensus as to the causes of the most recent rise in the Caspian Sea level. Most specialists believe that the main causes are long-term hydroclimatic factors, although several other hypotheses have been advanced. One hypothesis is that tectonic processes along the Caspian Sea coast have played a role in the observed increase. However, land movements have amounted to just a few millimetres per year, and in different directions in different parts of the coast, so that it is unlikely that such processes have been significant. According to another hypothesis (Shilo and Krivoshei, 1989), the compression of individual rock layers and folded structures have caused underground water to drain into the sea, raising the sea level, and expanding tensions have caused sea water to be drawn underground, thus lowering the sea level. This hypothesis is now considered to be highly unlikely, however.

Here we consider in more detail the hydrometeorological conditions in the basin during the most recent period of sea level rise. Table 7.2 presents data on river inflows to the Caspian Sea. In 1978–93 the average annual inflows were 307 km^3/year, which is 18 km^3/year above the long-term average of 289 km^3/year (1880–1993). In 1978–93 the share of inflows from the Volga (259 km^3/year) rose from 80% to 84% of total inflows; in 1978–93 the Volga discharges were 21 km^3/year higher than the long-term average (238 km^3/year), and 45 km^3/year higher than the 1933–77 average (214 km^3/year), when the sea level fell. The Volga inflows were particularly high in 1990 and 1991, when the sea level rose by a total of 73 cm.

Table 7.1. *Water balance components of the Caspian Sea since 1880.*

| Period | Sea level change (cm) | Inflows to the Caspian Sea[a] | | Outflow to Kara-Bogaz-Gol Gulf | | Visible evaporation (mm/yr) |
		(km^3/yr)	(cm/yr)	(km^3/yr)	(cm/yr)	
1880–1913	−92	306	75.5	27.4	6.7	715
1914–32	+16	321	80.4	19.7	4.9	747
1933–40	−172	229	58.5	10.5	2.7	773
1941–56	−61	292	77.5	11.6	3.1	782
1957–70	−4	281	75.5	9.5	2.6	732
1971–77	−61	236	65.2	6.9	1.9	720
1978–93	+229	312	82.8	4.5	1.1	670
1880–93	−145	294	75.3	16.6	4.2	735

[a]Including underground water inflows, equal to 5 km^3/yr.

Table 7.2. *River water inflows to the Caspian Sea, 1978–93 (km^3/yr).*

Years	Volga River	Ural River	Rivers from the Caucasus and Iran	Total inflows
1978	269	5.29	50.5	325
1979	297	6.24	34.8	338
1980	237	6.01	34.9	278
1981	280	9.24	35.1	324
1982	212	6.31	45.4	264
1983	220	9.46	35.2	265
1984	216	4.24	43.0	263
1985	285	9.08	28.6	323
1986	282	8.93	23.6	315
1987	263	11.2	40.5	315
1988	216	9.10	49.9	275
1989	214	6.84	35.9	257
1990	310	13.2	40.3	363
1991	307	13.2	41.0	361
1992	257	6.1	44.3	307
1993	270	17.0	46.8	334
Mean, 1978–93	259	8.8	39.4	307
Long-term mean, 1880–93	238	7.7	42.8	289

The Caspian Sea level is also affected by the relationship between precipitation and evaporation from the water surface, the so-called visible evaporation (evaporation minus precipitation). The data in Table 7.3 show that in 1978–93 visible evaporation was 670 mm/yr, 65 mm/year less than the long-term mean of 735 mm/yr. The evaporation from the water surface was 40 mm/year less than the mean, and annual precipitation over the sea increased by 20 mm/year.

The reduced evaporation during the last decade has been observed not only over the Caspian Sea surface, but also over much larger areas. From a recent study of evaporation from the water surface, using data provided by a special evaporation network, Golubev (SHI) concluded that evaporation has decreased systematically over the entire southern part of the former Soviet Union since the early 1970s.

Table 7.3. *Water balance components of the Caspian Sea, 1978–93.*

Years	Sea level change (cm)	Inflows to the Caspian Sea[a]		Outflow to Kara-Bogaz-Gol Gulf		Visible evaporation (mm/yr)
		(km³/yr)	(cm/yr)	(km³/yr)	(cm/yr)	
1978	19	330	92.4	5.7	1.6	718
1979	32	343	94.9	7.1	2.0	609
1980	11	283	77.9	0.0	0.0	669
1981	23	329	88.9	0.0	0.0	649
1982	3	269	72.3	0.0	0.0	693
1983	2	270	72.3	0.0	0.0	703
1984	3	268	71.4	1.0	0.3	681
1985	15	328	87.0	1.8	0.5	715
1986	7	320	84.8	2.0	0.5	773
1987	15	320	84.3	1.6	0.4	689
1988	15	280	73.3	1.6	0.4	579
1989	− 14	262	68.5	1.6	0.4	821
1990	37	368	96.2	1.6	0.4	588
1991	36	366	92.7	1.6	0.4	563
1992	15	312	78.4	12.7	3.2	602
1993	10	339	85.2	31.6	8.1	671
1978–93	229	312	82.8	4.4	1.1	670

[a]Including underground water inflows, equal to 5 km³/yr.

At the SHI estimates have been made of the effects on the sea level of the closure of the Kara-Bogaz-Gol Gulf in 1980–84, and the reduced flow from the sea into the gulf by up to 1–2 km³/year in 1985–90. This reduced considerably the flow of water into the gulf compared with conditions of free exchange, and led to a rise in sea level of 3–4 cm/year. The total volume of the sea in 1980–90 was increased by about 134 km³, leading to a rise of 35 cm, or about 15% of the most recent increase.

The rise in sea level since 1978 has therefore been due to hydroclimatic factors, which determine changes in the relations between elements of the water balance: increasing incoming terms such as surface inflows (mainly Volga runoff) and precipitation over the sea; and decreasing outgoing terms, such as evaporation from the water surface. The closure of the Kara-Bogaz-Gol Gulf also promoted an increase in the level. This analysis of the sea water balance in 1978–93 accords with independent assessments of the various components, so that the effects of other factors on level variations (tectonic, tensions in the Earth's crust, etc.), if any, are insignificant or within the limits of error in the determination of the water balance terms.

Figure 7.4 shows the relationship between the sea level and total river inflows (minus the outflow to the Kara-Bogaz-Gol Gulf) in 1978–93, and the most recent significant fall in sea level in 1930–40. With annual inflows of 340–365 km³, the sea level rises by more than 30 cm/year, and with annual inflows of 200–220 km³ it falls by 20–30 cm/year. A series of such water-abundant or water-scarce years has been the cause of the extreme level changes.

The main contributor to the increased inflows to the sea in recent years has been the runoff of the Volga, due largely to increased precipitation in the catchment, and possibly also to the impacts of human activities on river runoff. Table 7.4 shows the average annual precipitation recorded at 46 meteorological stations in the Volga catchment for the periods in which the most drastic changes in sea level occurred. In 1933–40 annual precipitation over some parts of the catchment (in the Upper Volga, and in the Kama and Oka basins) was 20–70 mm less than the long-term mean, whereas in 1978–90 it was 40–60 mm above the mean. Most of this latter increase has been in the winter precipitation, which contributes most to the total river runoff. These changes in precipitation have

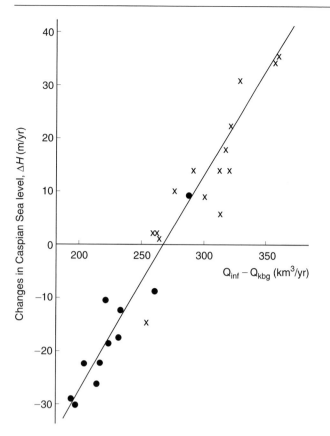

Figure 7.4 Relation between the changes in the Caspian Sea level (ΔH) and total river inflows (Q_{inf}) minus the outflow to the Kara-Bogaz-Gol Gulf (Q_{kbg}). • 1930–40; × 1978–93.

Table 7.4. *Average annual precipitation at various locations in the Upper Volga, Kama and Oka sub-basins of the Volga catchment (mm/yr).*

Region	1891–1990	1933–40	1978–90
Kalininskaya	595	572	637
Permskaya	547	478	585
Bashkirskaya	473	386	524
Kaluzhskaya	616	632	667
Tulskaya	548	532	608
Kostromskaya	580	534	623

inflow has had a significant effect on the water balance of the Caspian Sea.

Potential future sea level changes

The Caspian Sea level is highly sensitive to variations in the natural water balance components and to human activities within the basin. To obtain reliable estimates of the sea level in the future, the changes due to climatic factors and those due to human activities need to be distinguished. The greatest difficulties arise in identifying the former changes, particularly when attempting to forecast natural variations 10, 15 or more years in advance. This problem was addressed in the 1960–70s, and recent studies have attempted to forecast average annual levels over the next 5–30 years. The methods used to estimate future levels of the Caspian Sea are based on three basic principles:

- to establish relationships between the sea level and solar activity (usually expressed as Wolf numbers);
- to determine the relations between the variations in the sea level and atmospheric circulation patterns; and
- to determine the periodicity of long-term variations and to make extrapolations into the future.

The results of such forecasts are, however, often ambiguous and sometimes even contradictory (Shiklomanov, 1988; Kopaliani *et al.*, 1995). Predictions of the sea level over the next 10–20 years have ranged from a rise of 3–5 m to a fall of 2 m. Obviously, sound decisions on the most appropriate coastal protection measures can not be made on the basis of such unreliable results.

The methodologies used to determine future levels of the Caspian Sea form part of a larger effort to forecast long-term average climate conditions over large regions. The problem is far from being resolved, and the reliability of forecasts is very

been the main causes of the variations in Volga runoff, and the subsequent changes in sea level.

Preliminary analyses of the causes of the increased precipitation in the Volga basin show that there have been changes in atmospheric circulation patterns over European Russia, particularly in the trajectories of cyclones that bring moisture from the Atlantic. In the last 15 years they have shifted southward, increasing precipitation in the Volga catchment. The causes of these shifts are difficult to determine, and almost impossible to predict; they could represent natural long-term variations in climate, or could be the result of human-induced climate change. Human activities over the last decade may also have affected river runoff within the Caspian basin (irrigation, reservoirs, industrial or domestic water consumption, etc.). In the early 1980s water withdrawals within the basin amounted to 35–40 km³/year, but in recent years economic reconstruction in the former Soviet Union has led to sharp reductions in industrial and agricultural production (by up to 50%). Water withdrawals have decreased noticeably, by 10–15 km³/year, and the increased poor. Many researchers admit that the correlation coefficients between sea level and atmospheric circulation characteristics contain systematic errors and are unstable, and so do not allow accurate long-term forecasts. Another drawback is the lack of reliable methods for forecasting atmospheric circulation patterns, since they are often based on analogies, and are necessarily arbitrary because the trends in atmospheric processes in the past may not always continue in the future.

The solar-terrestrial relations used in long-term forecasting have no reliable physical or experimental basis, and their stability is poor. This has been confirmed in studies at the SHI of the correlation between the Caspian Sea level and indices of solar activity and atmospheric circulation (Georgievsky, 1978). The results of long-term forecasts of the Caspian level should be used with caution, even if they are based on relations with high correlation coefficients. Depending on the period chosen for the analysis and the factors to be analysed, it is possible to obtain quite different and even opposite inferences about future variations in the sea level. These shortcomings do not mean that there is no means of forecasting long-term changes for closed water bodies. The relations between the elements of sea water balance and spatial and geophysical factors are complex, and they are difficult to determine and interpret. The situation is worsened by the fact that future forecasts need to take into account the possible changes in global climate due to increased emissions of CO_2 and other 'greenhouse' gases.

Another approach to calculating the future sea level is based on the probabilistic nature of variations in hydrometeorological elements. At the SHI a mathematical model of the Caspian sea level was developed, with an equation for the sea water balance, and the Monte Carlo method was applied (Georgievsky and Ezhov, 1990). The model was used to estimate potential variations in sea level for two climate scenarios by the year 2020: stationary climate conditions, and non-stationary climate conditions due to global climate change. The model carries out joint modelling of the initial hydrometeorological series; calculates potential changes in river inflows to the sea due to human activities; estimates the effects of changes in evaporation and precipitation on the sea level, and on major river inflows (primarily the Volga) due to global climate change; and estimates potential outflows from the sea to the Kara-Bogaz-Gol Gulf.

In estimating the future sea level the model uses the following initial data:

- Probable future values of the sea water balance components for both stationary and non-stationary climate

scenarios based on calculations using actual long-term data.

- To estimate changes in inflows due to human activities it is assumed that with the recent reductions in industrial and agricultural production in the basins of rivers flowing to the Caspian Sea, water consumption would be lower. In the late 1980s the total river water losses due to abstraction and evaporation from reservoir surfaces were about $40 \, km^3/year$, and in 1995 these declined to about $32 \, km^3/year$. It is assumed that water consumption will remain at present levels until the year 2000, after which it will gradually increase again to $45 \, km^3/year$ by 2010.

- The dam closing the Kara-Bogaz-Gol Gulf was dismantled in mid-1992. In the subsequent three years about $150 \, km^3$ of water flowed into the gulf, and the level rose by $8 \, m$; it is now close to the sea level. The outflow from the sea into the gulf will be limited by the evaporation rate, corresponding to $18 \, km^3/year$. Subsequently, the outflow will depend on the sea level.

- Under the non-stationary climate scenario, estimates of expected changes in climate parameters in the Caspian basin (air temperature, precipitation) and in the water area (precipitation, evaporation) were taken from studies carried out at the SHI (Budyko), and potential changes in river inflows to the sea were obtained using a specially developed water balance model.

Calculations of the Caspian level for the period 1996–2020, using an initial sea level of $-26.6 \, m$, yielded the following results. Under the stationary climate scenario it is likely that by 2010 the sea level will have fallen by $1.1 \, m$, and by 2020 by $1.8 \, m$ (± 0.6–$0.7 \, m$). Under the non-stationary climate scenario, the sea level in 2005 is the same as under stationary conditions. Subsequently the level falls, but less rapidly, and after 2010 it begins to rise. The standard error for sea level changes only slightly with different variants of the calculations, and is determined almost entirely by the variations in the main water balance components of the sea.

Thus a very wide range of changes in sea level is possible, depending on future climatic conditions and human activities in the basin, which need to be taken into account in the long-term planning of large-scale water management projects. Of particular value is the future scenario of global climate change caused by increased emissions of CO_2 and other 'greenhouse' gases.

7.2.2 Russia, Siberia, Japan and China

The relations between climate and hydrological characteristics have been studied in Russia, Japan, China and elsewhere, based on observation data and analyses of hydrological and water regime characteristics over the long term, with changes in air temperature and precipitation within river basins or regions. In this way, observation data are used to assess the sensitivity of river systems to possible future changes in climate.

Russia

In recent years, hydrometeorological, hydrochemical and water management information archives have been compiled for river basins in Russia and elsewhere, and data on air temperature, precipitation and water regime characteristics have been analysed for several river basins in the northern hemisphere, including the Asian territory of Russia. From these analyses it has been found that in the last 10–15 years there have been significant increases in air temperature, particularly in winter, in most river basins in the northern hemisphere. In some regions (e.g. western and central areas of European Russia and eastern Siberia) the winter air temperatures have been the highest observed in the last 100 years, resulting in increased river runoff throughout European Russia.

Siberia

In eastern Siberia, the increased winter air temperatures have not resulted in significant changes in river runoff in winter. In a study of the Yenisey River basin in Siberia (Shiklomanov, 1994), nine sub-basins with long-term observation data and with undisturbed water regimes were selected, in each of which meteorological stations have recorded air temperature and precipitation data and monthly hydrometeorological data for more than 60 years. From these data it has been found that in some periods there have been significant deviations from the mean values in various parts of the basin. Since 1975 a positive air temperature anomaly has been traced over the whole basin, with increases in maximum air temperatures in winter of up to 2.5°C. Changes in river runoff, air temperatures and precipitation in the Yenisey basin in 1975–90 have been studied; the maximum warming was observed in the south (Upper Yenisey, Angara), and the minimum in the north, near the Arctic Ocean coast. According to global warming forecasts, this process should have been the reverse, i.e. the warming should have been

greater in the north. Precipitation has been higher throughout the basin, ranging from +25 mm in the Trans-Baikal and central regions, and up to +100 mm in the south and north of the Yenisey basin. The increased precipitation in the southeast occurred in the summer, and in other areas during the winter.

The changes in air temperature and precipitation have affected annual and seasonal river runoff in the basin only slightly, by ±4–6%, which is within the limits of natural long-term variations. Unlike in European Russia, the higher winter air temperatures have not resulted in increased river runoff because mean air temperature in this region range from −15 to −20°C, so that even an increase of 3–4°C would not cause the permafrost to thaw and produce more runoff in winter.

Japan

In Japan, long-term hydrometeorological data recorded over the past 100 years were analysed in a preliminary assessment of the effects of global warming on hydrological regimes. The study compared precipitation regimes and runoff characteristics during the warmest and coldest 10-year periods, when the mean annual air temperature difference was 0.74°C. It appeared that a 10% increase in precipitation in summer was due to more frequent storms (with more than 300 mm falling on just two days). Meanwhile, the total precipitation for 60 and 90 days with low rainfall in summer was much less than in winter (Yamada, 1989). This analysis gave some grounds for believing that increases in maximum precipitation and runoff during storm periods would occur in the case of global warming. In Japan this would cause problems with runoff control and flood protection in winter, and the reduced runoff during the dry season could worsen the existing water supply problem.

China

In an analysis of hydrometeorological data for northern China, Chunzhen (1989) showed that since 1981 air temperatures have been the highest for the observation period of 250 years. In 1981–87 the mean air temperature was 0.5°C above normal, and precipitation was below normal (−4% in Beijing), causing significant reductions in water resources. Based on a study of natural climate variations over the last 100 years, the authors assume that the warming that began in the 1980s in northern China will continue until the end of the century.

7.3 IMPACTS OF GLOBAL WARMING ON RIVER RUNOFF

7.3.1 Methodological approaches

In recent years there have been many investigations of the effects of climate change on hydrological characteristics and water resources in various regions and river basins in Australia, China, India, Indonesia, Japan, Malaysia, New Zealand, Russia and Thailand. In particular, in an analysis of possible hydrological regime changes in the Mekong River, one of the largest international water systems in the world, some attempts were made to estimate the effects of global warming on water availability and total water resources. Although the results of these studies are usually incompatible because they are based on different basic data, on different methodological approaches, and on different scenarios for the future, they nevertheless provide some indications of the sensitivity of river systems to changes in climate parameters. Also, general trends in the dynamics of water resources and water availability in the case of global warming can be identified.

Most of these investigations have focused on assessing the hydrological consequences of global warming under various climate scenarios. In some cases the relations between long-term climate parameters and hydrological characteristics have been analysed; only a few studies have dealt with specific problems of water resources management and water availability under conditions of climate change. This may be explained by the low rate of runoff control in the large river systems of Asia.

Most studies of Asia and Australia have attempted to assess possible changes in hydrological regimes due to global warming. Because reliable forecasts of regional climate variations are not available, various future climate scenarios have been applied: hypothetical scenarios, based on historical data; scenarios based on global circulation models, and paleoclimate reconstructions that assume changes in climate characteristics.

Hypothetical scenarios assume that air temperatures will rise by 0.5–4°C and precipitation will change by ±10–25%; some also specify changes in evaporation. Hypothetical scenarios are often based on long-term observations or historical data. The global warming expected in the next 30–50 years has not been witnessed during the whole of human history, so it is natural that with hypothetical scenarios unrelated to particular time periods in the future it is possible to assume the likely responses of river systems to variations in some climate parameters.

Global circulation models (GCMs) are usually used to compute climate characteristics in the case of a doubling of CO_2 concentrations $(2 \times CO_2)$ (Chang *et al.*, 1992). Of the various GCMs available, the IPCC (1990) recommended three models for assessing the regional impacts of global warming: the GCMs of the Geophysical Fluid Dynamics Laboratory (GFDL), the Canadian Climate Centre (CCC), and the UK Meteorological Office (UKMO). These models provide the highest spatial resolution and (in the authors' opinion) the most reliable results. Other GCM-type models have been developed by Australian meteorologists, and at the International Institute of Applied Systems Analysis (IIASA) additional scenarios have been used to assess the impacts of climate change on southeast Asia obtained from the GISS (Goddard Institute for Space Studies) and OSU (Oregon State University) models. The GISS GCM is most often applied in Asia and Australia. These models can be used to obtain detailed changes in regional climates (monthly and even daily air temperatures and precipitation) for almost the entire surface of the Earth. Unfortunately, the reliability of such computations is doubtful; the different GCMs often give variable and even contradictory results for the same regions (Kirshen and Fennessey, 1992). Scenarios based on GCM outputs may therefore be just as accurate as hypothetical scenarios, at least for precipitation.

Paleoclimate scenarios use climate conditions of the past as prototypes for the future. Approximate charts of air temperature and precipitation distributions over the northern hemisphere are drawn for these prototypes using various indirect methods; the charts are then used to assess the hydrological impacts of climate change. The most detailed charts have been prepared for the former Soviet Union (Budyko, 1988), where they are widely applied. Scenarios based on paleoclimate prototypes are also widely used in New Zealand.

The use of paleoclimatic prototypes appears promising because they show trends of possible changes in hydrological regimes and in water resources that could occur in specified time intervals in the future, including the next 10–15 years. However, these methods have some limitations due to, first, the conventions of the accepted analogies between climates in the remote past and future, and second, the lack of reliable paleoclimate data for many world regions. In the assessments of the possible impacts of climate change on hydrological regimes in the areas considered here, the following methods have been used:

- statistical relations between runoff and meteorological characteristics for observation periods;
- long-term water balance models;
- deterministic hydrological models of river basins over the short term.

The first two groups of methods are used to estimate variations in annual runoff, due to their simplicity and the small amount of basic data required, although the results are often unreliable. Simple regression coefficients obtained for the past should be transferred to the future with great care, and the methods can not be used to estimate changes in seasonal and monthly hydrological characteristics. Long-term water balance models have been used by Russian hydrologists to assess possible changes in total runoff from large river basins and from physiographic and economic regions of Asia under global warming.

The third group of methods, deterministic hydrological models of river basins, is most intensively used; such models are preferable for updated conditions. It is possible to investigate explicitly the cause and effect relations in the climate–water resources system, to evaluate basin responses to variations in climate characteristics, and to compute possible changes in runoff under different physiographic conditions in cases where regional climate forecasts are available, and to outline future water projects.

Various types of hydrological models are used to assess the effects of climate change on hydrological regimes; these include the well-known Stanford model or the unit hydrograph model, which have been used for many years for hydrological computations and forecasts, and specially developed models for studying the hydrological impacts of global warming within different physiographic features.

7.3.2 Possible changes in hydrological regimes and water resources

There have been several investigations of the effects of global warming on hydrological regimes and water resources for many regions and river basins in Asia and Australia. However, it is difficult to draw reliable conclusions as to expected changes in the hydrological characteristics of a particular region due to the wide variety of GCM-based climate scenarios, methodological approaches, and data required. All model assessments of future climate change are highly uncertain, particularly changes in precipitation. The results should therefore be regarded only as indicators of the responses of river systems to variations in climate characteristics, or, at least, as potential trends in relation to particular climate scenarios.

In the following, the results of assessments of possible hydrological impacts of global warming are presented for three physiographic and climatic regions of Asia: (1) cold and temperate; (2) arid and semi-arid; and (3) the humid tropics and monsoon regions.

(1) Cold and temperate regions

Various studies have been made of Siberia, the mountain regions of central Asia and New Zealand. In the river basins in the cold and temperate regions of Siberia the major proportion of annual runoff is derived from spring snowmelt. At the SHI studies have been made of variations in annual runoff in all river basins of the former Soviet Union. Scenarios for global warming by 1–2°C, based on paleoclimate analogies, were used as regional climate characteristics. The computations were based on the use of the long-term water balance method, and evapotranspiration was computed using the methodology developed and tested at the SHI. The results indicate that with a warming by 1°C, annual runoff can be expected to increase by 5–10% in the rivers of the Caucasus, Siberia and the mountain regions of central Asia. For mountainous areas, however, the assessments are approximate because of wide uncertainties in the paleoclimate scenarios. With global warming by 2°C, in almost all large river basins in Siberia and the Far East, annual runoff would increase by 10–20%. The same general trend can be expected in the river mountain basins in central Asia.

For the Yenisey River basin, Shiklomanov (1994) studied possible changes in annual, monthly and seasonal runoff using a special water balance hydrological model with 10-day time intervals. With warming by 1°C, annual runoff in the basin may be 14% higher, moreover, spring and summer–autumn runoff would be 11% higher and winter runoff 15% higher. With even greater global warming the trend towards higher winter runoff and lower spring runoff would remain. Detailed results are given in the Yenisey River basin case study (see below).

The results of studies of the rivers in the cold and temperate regions of Siberia have generally confirmed the conclusions drawn for the rivers of Russia and elsewhere in Europe where the major portion of river runoff is derived from spring snowmelt (Shiklomanov and Lins, 1991). In these regions seasonal runoff and the streamflow distribution during the year, unlike annual runoff, appear to be more sensitive to changes in air temperature changes than in annual precipitation. Estimates of changes in future seasonal runoff appear to be more reliable than those of annual runoff because forecasts of air temperatures based on GCMs are always more reliable than forecasts of precipitation.

For New Zealand, the variations in water resources with global warming have also been assessed on the basis of a paleoclimate scenario. About 8000–10 000 years ago, mean air temperatures were 1.5°C higher than today. From maps of isolines of air temperature and precipitation for that period, Salinger and Hicks (1989) found that air temperatures over most of the country were 1.4–1.6°C higher and precipitation varied within − 10% and + 15%; moreover, precipitation decreased over 20% of the country. The changes in annual runoff, estimated using a long-term water balance method, are as follows (Griffiths, 1989):

- North Island: 9–40% decrease in annual runoff in the southeast, and 9–27% increase over the rest of the island (80–90% of the area);

- South Island: 18–40% decrease in runoff in the southeast; over the rest of the island (70–75% of the area) runoff is unchanged or tends to increase by 6–40%.

- Together with significant changes in annual river runoff over most of New Zealand, it is expected that with global warming by 1.5°C the frequency and rates of rainfall floods would increase.

Miller and Russell (1992) used the GISS with a $2 \times CO_2$ scenario to assess potential changes in annual runoff for 33 of the world's major rivers. For the Yenisey, Lena, Ob, Amur and Kolyma, the authors found significant increases of 10–45% in annual runoff. In a recent study of the Yenisey basin using paleoclimatic scenarios, Shiklomanov (1994) found considerable increases in runoff with global warming (see the case study below).

CASE STUDY: *Effects of global warming on the Yenisey River basin*

The Yenisey River in Siberia is the largest river system in Russia, with a mean annual discharge of 630 km^3/year. The river rises in the mountains and semi-deserts of northern China and Mongolia, and flows through taiga forest and tundra to the Kara Sea of the Arctic Ocean. The basin extends over an area of 2.58 million km^2; the climate is continental, varying from temperate to cold.

Since the 1960s five large dams have been constructed on the Yenisey and its tributary the Angara for long-term water storage and seasonal runoff control. The total volume of the reservoirs is 333 km^3, with an effective volume of 96.5 km^3. The dams were also intended to generate power for the industrial centres of Krasnoyarsk, Irkutsk, Bratsk and Ust-Ilimsk. About 300 000 hectares in the southern part of the basin are irrigated.

To evaluate the effects of global warming on the hydrological regime of the basin, three climate scenarios were developed at the SHI (Budyko, 1988), based on paleoclimate reconstructions of past warm periods. The results of these scenarios were output in the form of maps of isolines of air temperatures in summer and winter, and total annual precipitation.

Scenario 1: Warming by 1°C by the year 2000. Average annual air temperatures over the whole basin can be expected to rise by 1.5–2°C. Winter air temperatures are expected to rise by 1.5–2°C in the

south and southeast, and by 3–4°C in the centre and north of the basin. Summer air temperatures would show little change. Annual precipitation can be expected to increase by an average of 15%, by 50 mm in the south and centre, and by up to 90 mm in the north of the basin.

Scenario 2: Warming by 2°C by the year 2020. Winter air temperatures would rise by 6–7°C in the south and centre of the basin, and by up to 9°C in the north; summer air temperatures would rise by 2–4°C. Changes in precipitation would be similar to those under scenario 1.

Scenario 3: Warming by 3–4°C by the year 2050. Air temperatures would rise by 8–9°C in the south, by 10–12°C in the centre and southeast of the basin, and by 16–18°C in the north. Changes in annual precipitation would vary over the basin, with an average increase of 110–120 mm/year.

In fact, scenario 3 includes scenario 2, and scenario 2 includes scenario 1. The temperature rises calculated for the Yenisey basin are surprisingly high. Such significant changes in meteorological parameters would have serious effects on the runoff characteristics in the Yenisey basin. Scientists at the SHI therefore developed a water balance

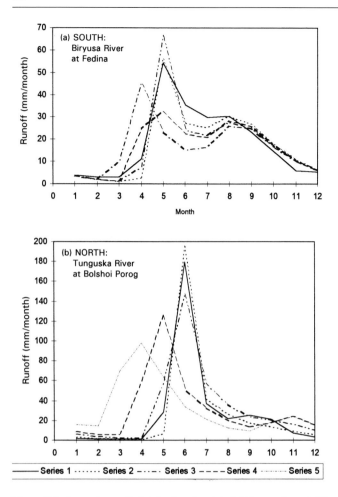

Table 7.5. *Annual runoff in the Yenisey River under the three global warming scenarios (km³).*

Climate	Spring	Summer–Autumn	Winter	Year
Contemporary	396	144	90	630
Scenario 1	428	166	126	720
Scenario 2	422	156	141	719
Scenario	445	168	160	773

Figure 7.5 Annual runoff distribution (mm) in (a) the north, and (b) the south of the Yenisey basin. 1, observed runoff; 2, model results; 3–5, global warmig by 1, 2 and 3–4°C, respectively.

model, with 10-day time intervals, to evaluate the effects of global warming on the regimes of the Yenisey basin rivers. Existing data on air temperature, precipitation and humidity were used to compute the hydrological cycle components: evaporation, soil moisture, surface and subsurface runoff. The model is based on the joint solution of energy and water balance equations and is intended for a continuous determination of the water balance components. The model was developed as follows:

1. Current climate conditions were characterized using long-term data on precipitation, air temperature and humidity, recorded in each of six representative sub-basins.
2. Model parameters were derived from analyses of published data, soil maps, etc.
3. The model was run to compute mean water balance components for conventionally natural periods (up to 1974) from observed meteorological data.

4. The accuracy of the model was assessed by comparing actual and computed water balance components. Figure 7.5 compares two such hydrographs for the south and north of the basin; the mean relative error is less than 15%. It was concluded that the model provided reliable descriptions of the process of runoff formation in the basin and that it could be applied to assess possible changes in runoff due to climate change.
5. According to the regional climate change scenario, appropriate corrections were made to the observed air temperature and precipitation data; these were then used to compute the water balance components.
6. Possible changes in runoff were obtained by comparing actual and computed values.

The three climate change scenarios were used to obtain runoff hydrographs and inter-annual distributions of water balance components. The seasonal streamflow distribution is shown in Table 7.5. The main results were as follows:

Scenario 1: Significant increases in annual runoff can be expected, ranging from 20–25 mm in the south, to 55–60 mm in the north of the basin. Runoff would increase throughout the year, and the streamflow distribution would remain unchanged. Surface evaporation would increase by 15–35 mm/year. In the event of a particular combination of the climate conditions, the annual runoff volume of the Yenisey could increase by 90 km³, most of which would occur in winter and spring.

Scenario 2: Annual runoff would be similar to that under scenario 1, but the streamflow distribution would change significantly, particularly in the north and centre of the basin, where spring snowmelt floods would occur one month earlier. In the south the spring flood maximum would be much lower, but the peak would be similar to that under scenario 1 and a seasonal streamflow redistribution would occur.

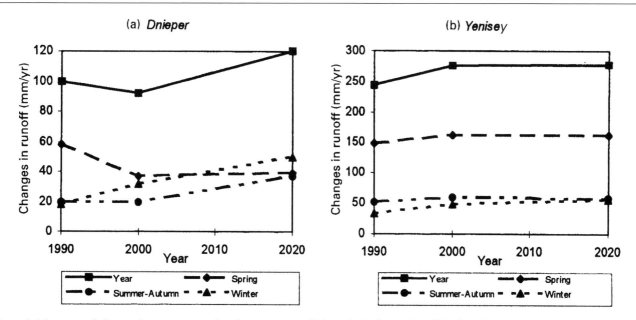

Figure 7.6 Impacts of climate change on annual and seasonal runoff from the Dnieper (a) and Yenisey (b) rivers. Paleoclimate scenarios for 2000 and 2020.

Scenario 3: In this (least probable) scenario, annual river runoff would increase by 50 mm in the south, and by up to 70–75 mm in the north. Spring snowmelt floods could occur up to two months earlier, and would last longer. Annual runoff would increase by 23%.

Figure 7.6 compares the changes in annual and seasonal runoff of the Dnieper and Yenisey rivers according to these three scenarios. For the Dnieper, as well as other rivers in regions with mild winters, changes in seasonal runoff could be significant due to increased winter runoff and reduced spring runoff due to more intensive snowmelt in winter (Shiklomanov and Lins, 1991). In the Yenisey basin, where the climate is more severe, the changes in streamflow distribution during the year are not significant (see Figure 7.6), since streamflow is dependent on precipitation rather than air temperature.

The preliminary conclusions drawn from this case study were as follows:

- the observed increases in air temperature and precipitation in the Yenisey basin since 1975 have indirectly confirmed that global warming is occurring;
- the rates of increase in air temperature and precipitation in the region exceed current mean global values;
- the recent increases in mean air temperature and precipitation have not led to annual and seasonal variations in runoff;
- according to all climate change scenarios, runoff in the basin would increase, the seasonal runoff distribution would change, and the winters would be shorter.

Further studies are necessary to take into account the effects of global warming, by 2–3°C in particular, on the permafrost zone, and on the hydrological regimes of the Yenisey basin rivers.

(2) *Arid and semi-arid regions*

According to Chunzhen (1989), even minor changes in climate characteristics would have a major impact on the hydrology of river basins of northern China. The results of a special hydrological model showed that in semi-arid regions, assuming a 10% increase in precipitation and a 4% reduction in evaporation, runoff would increase by 27%. With a 10% increase in precipitation and a 4%

increase in evaporation, runoff would increase by 18%. Similar changes in climate parameters in more arid regions would cause increases in runoff of up to 30–50%.

For Australia, studies of the hydrological impacts of global warming on the arid and semi-arid regions have been under way since 1988. In 1991 the results and their implications were discussed at a special meeting on hydrology and water resources, organized by the Australian Bureau of Meteorology (1991).

Detailed studies have been made of the effects of climate change on hydrological characteristics and water resources in the Murray–Darling basin, which drains an area of over 1 million km^2. The mean annual runoff is 12.4 km^3, of which 10 km^3 is controlled. Average annual water consumption exceeds 8.6 km^3, or 60% of the regional total. The basin is an important agricultural region (Zillman, 1989), accounting for 75% of the irrigated land area of Australia, and is of major socio-economic importance. It is also considered to be extremely vulnerable to climate variations. In a study of the behaviour of the Murray River system over the last 94 years, a special hydrological model was developed to assess the influence of global warming on water resources and existing water control systems (Close, 1988). The conclusions were as follows:

- runoff from tributaries throughout the Murray–Darling basin would increase;
- the increase in river runoff in the north of the basin would be greater than in the south;
- precipitation in spring, autumn and in summer would increase, while that in winter would be slightly lower;
- the specific (per hectare) irrigation water requirements would be somewhat lower;
- more water would be available for irrigation, allowing either an increase in the irrigated land area, or improving water supplies for existing irrigated lands;
- runoff capacity for dilution would be greater, thus reducing the rate of salinization downstream.

In a study of Australia's underground water resources, Ghassemi et al. (1991) confirmed the above conclusions. They noted that precipitation has changed significantly throughout this century, and that this is likely to continue in the case of global warming. They concluded that for most of Australia, particularly the densely populated southeast, where underground water resources have been exhausted, the rate of recharge of underground water would increase. Shallow aquifers in arid and semi-arid areas would benefit from the higher precipitation and infiltration. However, for some important regional aquifers in the southwest, global warming would probably result in an unfavourable reduction in precipitation.

These optimistic assessments of the impacts of global warming in Australia were confirmed by Budyko (1988) using paleoclimatic reconstructions. Global warming will probably result in increased precipitation over most of the continent. However, together with greater water availability, some authors expect greater variability in the extreme characteristics of precipitation and river runoff (Australian Bureau of Meteorology, 1991). Model results with a $2 \times CO_2$ scenario show that maximum precipitation tends to increase in summer and winter, while minimum precipitation tends to decrease. These trends are observed almost everywhere in Australia, even where total annual precipitation is expected to decrease.

For both China and Australia, global warming is expected to result in increases in the severity and frequency of floods (Liu et al., 1995). Bates et al. (1995) used a stochastic weather generator coupled with two daily rainfall runoff models to investigate changes in the behaviour of annual maximum monthly runoff series for six Australian catchments within a variety of climatic settings. In five cases the series were noticeably higher for a changed climate than for the present day.

(3) Humid tropics and monsoon regions

This subsection describes the results of studies of the impacts of global warming on hydrological regimes and water resources in the humid tropics and monsoon regions of the Mekong River basin, southeast Asia (Indonesia, Malaysia and Thailand), and regions of Sri Lanka and India.

The Mekong River basin
About 75% of the area drained by the Mekong River (the largest international basin in Asia), lies in the tropics, and has a typical seasonal monsoon climate. In the studies of possible water resource variations (Mekong Secretariat, 1990) three GCMs (GISS, GFDL and UKMO) were used as the basis of climate scenarios. For all hydrological stations the three GCMs showed increases in the duration of both the dry and wet periods. The monsoon season is expected to occur one or two months earlier, together with changes in the timing of the beginning of the driest and wettest months. In general, precipitation during the wet months tends to increase, with a slight decrease in rainfall during the dry months, and the duration of periods with rainfall deficit tends to be longer. Changes in the control of the outflow of reservoirs will therefore be necessary, in order to adjust to the greater variability periods of high and low flows. Such new runoff control regimes are likely to affect power generation and irrigation to some extent, and may lead to undesirable consequences.

Southeast Asia
In the early 1990s, UNEP (Parry et al., 1991) and IIASA (Toth, 1993) attempted to evaluate the hydrological impacts of global warming on various watersheds in Indonesia, Malaysia, and Thailand, and the socio-economic and political repercussions. These issues are of major concern in south-

east Asia, especially in regions with large differences between wet and dry seasons. The main conclusions of the $2 \times CO_2$ climate scenarios based on three GCMs (GISS, GFDL and OSU), were that precipitation during the wet season would increase, and that during the dry season the water deficit would be greater for several months. This would cause considerable problems for state authorities and institutions responsible for water management.

The most detailed studies of hydrological regimes in southeast Asia have focused on the Kelantan River basin in Malaysia. Annual precipitation in the basin is high (2200–3000 mm), up to 50% of which falls during the monsoon season (October–December). The monsoons cause severe floods, affecting both individuals and the national economy. During the floods of 1967, for example, more than 300 000 hectares were inundated, or 20% of the area of the basin. More than half a million people were affected, of whom 30 died and 125 000 had to be evacuated. The total cost of the flood damage exceeded $30 million. In contrast, during the dry season the water deficit affects agriculture, power generation, and industrial and domestic water supplies.

To evaluate the effects of global warming caused by a doubling of CO_2, the storage function model was applied to the flood flow of the Kelantan River, and the Thornthwaite water balance model was applied to study the monthly balance of inflow and outflow and on water resources. These two models were calibrated on the basis of recent observation data, and then, to make appropriate assessments, the climate parameters were changed on the basis of the GISS model. The results showed that the duration of floods would change slightly, and that flood peaks would be 9% higher, causing larger areas to be flooded. It was also found that the frequency of floods would increase from the present once in 50 years to once in 30 years.

The expected changes in the water balance components (runoff, evaporation and water deficit) for the Kelantan basin are shown in Figures 7.7–7.9 (Parry *et al.*, 1991). The most serious change is the increase in water deficit by 30–35% during the dry season; this would affect water supplies for irrigation, which are already critical.

The Thornthwaite water balance model was also used to assess the effects of climate variations on the water resources of three river basins in Indonesia (Toth, 1993). It was found that monthly precipitation would increase by 7–33% in the Citarum basin in western Java, by 5–50% in the Brantas basin in eastern Java, and by 8–59% in the Saddan basin in Sulawesi, and that higher air temperatures would affect the seasonal deficit and soil moisture content. The soil moisture content would increase in all three basins, and the present deficit would be eliminated completely in the Saddan basin.

In an additional study of the effects on runoff in the three basins, Rozari *et al.* (1990) concluded that monthly runoff would increase significantly, resulting in greater soil erosion and lower soil productivity, ultimately reducing crop yields by about 9%.

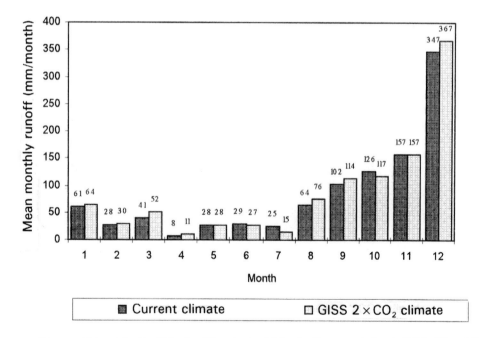

Figure 7.7 Mean monthly runoff (mm) for the Kelantan River basin, Malaysia, for current climate (1951–75) and the GISS $2 \times CO_2$ climate scenario (Parry *et al.*, 1991).

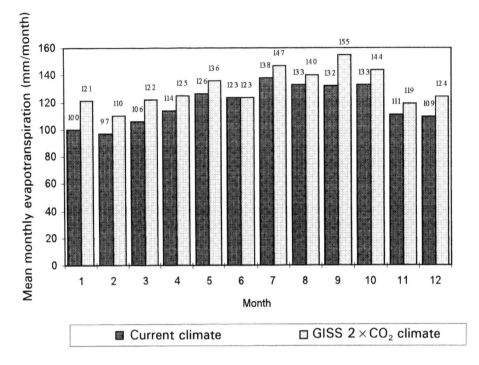

Figure 7.8 Mean monthly evapotranspiration (mm) for the Kelantan River basin, Malaysia, for current climate (1951–75) and the GISS $2 \times CO_2$ climate scenario (Parry *et al.*, 1991).

Sri Lanka

In a study of the effects of global warming on Sri Lanka, Nophadol and Hemanth (1992) compared the outputs of three GCMs (GISS, UKMO and GFDL) from the viewpoint of their ability to reproduce historical air temperature and precipitation data. They concluded that the GFDL model gave the least accurate results, and so applied the GISS and UKMO models for all estimates of the effects of a doubling of CO_2. The results showed that there would be little change in the mean annual precipitation, but mean monthly

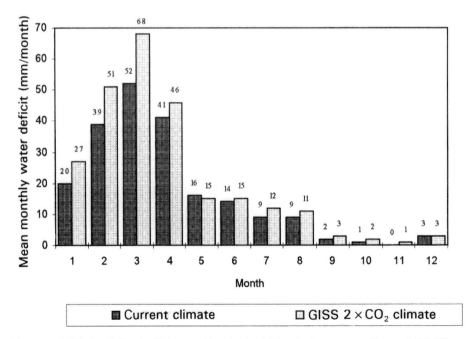

Figure 7.9 Mean monthly water deficit (mm) for the Kelantan River basin, Malaysia, for current climate (1951–75) and the GISS $2 \times CO_2$ climate scenario (Parry *et al.*, 1991).

and daily precipitation would increase during the rainy season, and the maximum precipitation period would shift from April–July to September–November. Some danger of severe droughts during the low water period would remain, however.

India

For India, Leichenko (1993) examined the effects of global warming on the water supplies of the cities of Bombay and Madras. These densely populated areas were selected because their water supplies depend directly on precipitation during the rainy season, and are likely to be vulnerable to changes in air temperature and precipitation. These cities are already seriously affected by water deficits during droughts.

Both cities are growing rapidly; by the year 2050 the population of Bombay will have tripled from 13 to 44 million, and that of Madras from 6 to 20 million. Even now, the water availability in these cities does not satisfy the demand. For example, Bombay can provide 3 billion litres of water per day, whereas the demand is about 3.5 billion l/day. The situation is even worse in Madras, where the demand for water exceeds the supply by 40%. Over the next 60 years water demand is likely to triple, even under stable climate conditions. The two cities also suffer from the variability of the monsoon climate, which causes periodic severe droughts or high floods. In the last 114 years (1871–1984) there have been 12 severe droughts and 10 floods in Bombay, and 17 droughts and 18 floods in Madras. Climate variability is a cause for concern for the water management and planning authorities, particularly because water storage during the monsoon is impossible due to insufficient capacity.

The possible effects of global warming on water resources are therefore of great concern for these two urban regions of India. The outputs of three GCMs (GISS, GFDL and UKMO) have been applied to develop $2 \times CO_2$ scenarios, although there are some drawbacks in the use of these models for conditions in India:

1. There is a large discrepancy between the resolution of the models ($500 \times 500 \, km^2$ grids) and the size of hydrological regions in India.
2. The models only poorly reflect the contemporary climates of many world regions. For India, for example, the $1 \times CO_2$ scenarios based on the GFDL and GISS models provide lower air temperatures than those observed, whereas those based on the UKMO model give much higher values. With all three models, the precipitation values for central and south India are overestimated (Kalkstein, 1991), and poorly reflect the contrast in precipitation between wet and dry seasons.

3. The models inadequately reflect some key processes, such as ocean circulation and cloud cover.
4. The existing models can not reliably predict the effects of increased CO_2 concentrations on climate variability; this is extremely important in assessments of water supplies in the monsoon regions of India.

Despite these shortcomings, these GCMs have been used to assess the impacts of global warming on future water resources in the Bombay and Madras regions. Further studies are needed of the effects of climate change on the timing and the intensity of the monsoon.

The results of the $2 \times CO_2$ scenarios based on the three GCMs vary widely, as can be seen in Table 7.6. For both cities the GFDL model gives the coldest and wettest scenarios, and the GISS model the warmest and driest. There is great uncertainty in the estimates, particularly for precipitation; the changes range from -20% to $+136\%$ for Bombay and from 0 to $+153\%$ for Madras. The UKMO-based scenarios are intermediate, and are probably the most realistic.

The effects of climate change on the hydrology of the Bombay and Madras regions have been studied by the water balance method using the aridity index developed by Oldekop (1911) and improved by Budyko (1948). The aridity index is derived from the ratio of evapotranspiration to precipitation, and is inversely correlated with mean runoff. Again, the $2 \times CO_2$ scenarios based on the three GCMs give widely differing results. The GFDL-based scenario gives a lower aridity index for Bombay (by 2–3 times) for the four monsoon months, which shows a sudden increase in runoff. The UKMO-based scenario gives slight changes in the aridity index for June–August and an almost doubled value in September. The GISS-based scenario also shows a sudden increase in the aridity index for the two wettest months, August and September. According to the last two scenarios, water availability in Bombay would be reduced during these months.

For the Madras region, the GFDL-based scenario gives a sudden increase in runoff; the UKMO-based scenario minor changes in runoff in September–November and a significant decrease in December; and the GISS-based scenario gives a decrease in runoff in November–December. Thus, the expected runoff decrease during usually wet months in this region, as well as in the Bombay region, would make the problem of water availability even more acute. The main conclusion from these scenarios is that there is great uncertainty of the future water availability in two of the largest and rapidly developing urban centres of India in the event of global warming.

Table 7.6. *Comparison of results of the three GCMs for the* $2 \times CO_2$ *scenarios for Bombay and Madras.*

		GFDL	GISS	UKMO
Bombay	Air temperature (°C)	+ 2.2	+ 3.7	+ 3.5
	Precipitation increase (%)	+ 136	− 20	+ 40
Madras	Air temperature (°C)	+ 1.3	+ 3.6	+ 4.2
	Precipitation increase (%)	+ 153	0	+ 10

7.3.3 General remarks

Despite the uncertainties in the assessments of the effects of global warming on the water resources of different regions and river basins of Asia and Australia, the general conclusion is that total water resources will tend to increase. Paleoclimate reconstructions indicate trends toward higher mean runoff in the rivers of Siberia and New Zealand, and in the areas of water deficit of Australia. Similar trends have also been observed in GCM-based scenarios for the tropical and monsoon regions of southeast Asia and for India. Miller and Russell (1992) found considerable increases in the total river runoff in Asia and Australia. For 10 of the 12 largest river systems of Asia and Australia annual runoff can be expected to increase by 10–48%, and for the other two (the Tigris–Euphrates and the Indus) runoff is found to decrease.

These results have been confirmed at the SHI (Shiklomanov and Babkin, 1992) using a paleoclimate scenario for the total water resources of all continents, and the long-term water balance method. With global warming by 1–2°C, total runoff in Asia will increase by 20–30%, and in Australia and Oceania by 10–18%. Although these estimates are based on just one paleoclimate scenario, in general they do not contradict the results based on the GCMs.

7.4 THE GEWEX ASIAN MONSOON EXPERIMENT (GAME)

7.4.1 Introduction

The Asian–Australian monsoon region extends from the tropical Pacific and Indian Oceans to the Arctic of Siberia and central Asia. In this region lives more than 60% of the world's population. The droughts and floods connected with the variability of the monsoon affect every aspect of human activity in the region. Global warming, and the resultant changes in water and the processes of heat exchange are

likely to increase this variability, with potentially adverse consequences for agriculture and the economic development of the region.

There are wide differences in model results for this region, indicating the considerable uncertainties in forecasts of changes in climate and hydrological characteristics, and the need for further research to characterize energy and hydrological cycles. In the early 1990s, within the framework of the World Climate Research Programme (WCRP), the Global Energy and Water Cycle Experiment (GEWEX) was established to study the basic physical processes related to the components of the global climate system, on time scales of one week to several years.

The Asian–Australian monsoon plays an important role in the global transport of energy and water. The Japanese National Committee of the WCRP initiated the idea of organizing within GEWEX a long-term international experiment to study the Asian monsoon: the GAME project. The first version of the scientific programme of GAME was outlined in January 1993, and was later discussed at various meetings and at two GAME conferences in China in 1994 and in Thailand in 1995. The main scientific tasks of the project have now been defined, as well as approaches and areas of investigation, and collaboration with other international projects. A detailed programme and work schedule are now being prepared (Yasunari, 1996).

7.4.2 The aims and tasks of GAME

The scientific aims of the GAME project are to improve understanding of the role of the monsoon in global energy and hydrological cycles; to improve the modelling and seasonal predictions of the monsoon using GCMs and numerical forecasting models; to improve understanding of multiscale interactions in the energy and hydrological cycles; and to evaluate the influence of the monsoon on hydrological cycles. The GAME project will focus on the following:

1. Observations of the energy and water cycles of the Asian monsoon system. It is planned to carry out continental-scale generalization of the observation data obtained from the ground gauge network, using geostationary satellite data over several years.

2. Observations of the heat balance on land and in the atmosphere in the Asian monsoon region, on the basis of combined ground-based and satellite data.

3. Macro-scale field investigations of the land–atmosphere interaction. First, satellite data will be used to determine all possible surface hydrological characteristics necessary for the parametrization of hydrological models

(snow cover, soil moisture, etc.), especially for the permafrost zone. Second, simultaneous detailed observations of river watersheds in different physiographic regions will be made.

4. Modelling and field-based studies of meso-scale convective systems and regional hydrological cycles. Radar observations will be used in combination with detailed field investigations and data obtained in (3), in order to elaborate heat–water exchange models with a horizontal grid size of about 1×1 km.

5. Studies of monsoons using coupled land/atmosphere/ocean system models; this will involve improving and using GCMs.

6. The four-dimensional data assimilation will require the generalization of all the programme results to create a reliable forecasting model with a horizontal resolution of 50 km and to improve the modelling of the effects of clouds and underlying surface parametrization.

7.4.3 The GAME programme structure

The GAME programme comprises the following elements:

(a) *Satellite observations.* GAME will make use of the satellites of other international programmes, such as GEWEX, TRMM (Tropical Rainfall Measuring Mission satellite), etc., and will also launch its own satellites to obtain high-resolution information on the land surface, soil moisture, snow cover and other hydrological characteristics in different physiographic regions.

(b) *Studies of processes and regional field experiments.* Intensive field observations will begin in 1998, followed by river regime observations up to 2005. GAME will focus on four regions of Asia: (i) tropical rainforests of southeast Asia: the Chao Praya basin, Thailand; (ii) the subtropical zone: the Huang He basin, China; (iii) the Tibetan high plateau (permafrost zone): the upper Yangtze basin, China; and (iv) the permafrost zone, taiga and tundra: the Lena basin, Russia.

(c) *Model studies.* Models will be improved and elaborated to provide seasonal estimates of the Asian monsoon, cloud systems, and hydrological processes. To integrate the satellite and field observation data, a comprehensive database will be compiled to enable their four-dimensional representation in a GCM.

(d) *Data collection and management.* The ground-based data will be used to create an extensive data-base of information for the GEWEX network. To meet the needs of all investigators, the database will include historical data, as well as ground and satellite observation data on various elements of the water and energy cycles, and their determining factors. The data will be organized as observation data (level I), information on geophysical parameters based on level I data (level II), and the data obtained from models or other special techniques (level III). The first draft of a data management plan for GAME is now being considered.

7.5 CONCLUSIONS AND RECOMMENDATIONS

This impressive range of research into the effects of climate variations on hydrological characteristics and water resources in Asia and Australia, can be summarized as follows:

- Asia and Australia are characterized by variable physiography, rapidly increasing populations, intensive human activities, and a high level of water resources development. Many regions are already facing problems due to water deficits, droughts or severe floods. Population growth will make all of these problems more acute in the future, so that reliable assessments of possible changes in climate and in water resources due to global warming are urgently required.

- For some regions and river basins, long-term hydrometeorological observations show that (1) in the past, even slight warming has caused significant changes in precipitation and river runoff regimes; and (2) in the last 10–15 years higher air temperatures and precipitation have been observed to have considerable effects on the hydrological characteristics and water resources in many regions.

- The results of GCM-based $2 \times CO_2$ scenarios and of paleoclimate scenarios for warming by 1–1.5, 2 and 3°C are used to assess the possible effects of global warming. The most widely used GCMs are the GISS and GFDL models (USA) and the UKMO model (UK).

- Other models that are used to study the hydrological impacts of climate change include river basin models (to assess changes in flood and low-water regimes) and annual and monthly water balance models, using the Thornthwaite method (to compute evaporation) or the

SHI method (to estimate changes in mean annual or seasonal runoff, and water resources deficits).

- In cold and temperate regions, where the major proportion of runoff is derived from spring snowmelt, global warming could lead to considerable changes in seasonal streamflow distributions. Changes in air temperature are likely to have a greater effect on runoff regimes than changes in precipitation, so that assessments of changes in seasonal runoff may be more reliable than those of changes in annual runoff and total water resources.

- In arid and semi-arid regions, river systems are most sensitive to changes in climate characteristics. Increases in precipitation by 10–20% and in air temperature by 1–2°C could result in increases in runoff by 40–70%.

- In humid tropical regions, runoff regimes depend mainly on precipitation; in most cases global warming would increase precipitation and runoff during the wet season, with less significant changes during the dry season.

- In different physiographic regions (from arid to tropical regions) global warming could lead to significant variations in seasonal and annual streamflow distributions, with increased maximum and reduced minimum discharges.

- The results of both GCM-based scenarios and paleoclimate reconstructions indicate that global warming will lead to increases in total water resources throughout Asia and Australia.

- The results of GCMs for monsoon regions are often unsatisfactory, and estimates of changes in precipitation in some regions obtained from different models are often contradictory. Further work is therefore required to improve the reliability of scenarios of the future climate.

- Further studies are required to assess the possible impacts of global warming: on hydrological regimes in arid and semi-arid regions of southwestern and central Asia and in mountainous regions; on water quality, sediment yields, and channel processes, as well as on systems of runoff control, water resources control, water availability and consumption; and, in particular, how economic and environmental systems can be adapted to cope with variations in hydrological regimes and water resources.

8 Overview of models for use in the evaluation of the impacts of climate change on hydrology

G. H. LEAVESLEY

8.1 INTRODUCTION

Atmospheric and hydrological models provide a framework within which the relationships between climate and water resources can be conceptualized and investigated. These investigations can range from simple one-way couplings using atmospheric model outputs as adjustment factors for measured inputs to hydrological models, to more complex, fully coupled atmospheric and hydrological model applications that incorporate feedback mechanisms between the two systems.

The scientific literature contains a large number of reports on the variety of atmospheric and hydrological models that have been used to investigate the effects of increasing greenhouse gases on climate, and the resulting impacts of any associated changes in climate on local and regional hydrology. Current atmospheric models have generally been shown to have a limited ability to simulate present climate conditions, and thus a large degree of uncertainty exists as to their applicability for impact assessments. Similar uncertainties have been identified in the applicability of the wide range of hydrological models. The purpose of this chapter is not to provide an extensive review of the findings of these reports, but to characterize the state of atmospheric and hydrological modelling for use in simulating the effects of climate change and current climate variability. Methodologies of modelling and climate scenario development are reviewed, their deficiencies are discussed, and additional research needs are identified.

8.2 HYDROLOGICAL MODELS

'Hydrological modelling is concerned with the accurate prediction of the partitioning of water among the various pathways of the hydrological cycle' (Dooge, 1992). This partitioning in its simplest form is expressed by the water balance equation:

$$Q = P - ET \pm \Delta S, \tag{8.1}$$

where Q is runoff, P is precipitation, ET is evapotranspiration, and ΔS is the change in system storage. Equation (8.1) is common to all hydrological models. The variety and number of hydrological models developed to solve it reflect the wide range of modelling purposes, data constraints, and spatial and temporal scales that have influenced the conceptualization and parameterization of the processes in the equation.

Models can be classified using a number of different schemes (Woolhiser and Brakensiek, 1982; Becker and Serban, 1990; Dooge, 1992). Classification criteria include the purpose of the model application (e.g. real-time application, long-term prediction, or process understanding); model structure (models based on the fundamental laws of physics, conceptual models reflecting these laws in a simplified approximate manner, black-box or empirical analysis); spatial discretization (lumped or distributed parameter); temporal scale (hourly, daily, monthly, annual); and spatial scale (point, field, basin, regional or global). A variety of these types of models have been applied to the assessment of the effects of climate change. This section reviews some of these modelling approaches in order to provide some measure of the state of the art of hydrological modelling. The following is a review of selected models to indicate the variety of modelling approaches and range of applications. The models are grouped by model structure and spatial discretization criteria.

8.2.1 Approaches

Empirical models

An empirical representation of Equation (8.1) considers only the statistical relations among the components of the water balance. An empirical model developed by Langbein *et al.* (1949) that expressed the relationship among mean annual precipitation, temperature and runoff was used by Stockton and Boggess (1979) to estimate changes in the average annual runoff of 18 designated regions throughout the United States for different climate change scenarios. Revelle and Waggoner (1983) used the same model to investigate the effects of climate change on runoff in the western United States.

Empirical models do not explicitly consider the governing physical laws of the processes involved, but only relate input to output through some transform function. As such, empirical models reflect only the relations between inputs and outputs for the climate and basin conditions during the time period in which these models were developed. Extension of these relations to climate or basin conditions different from those used for development of the transform function is questionable.

Water balance models

Water balance models originated with the work of Thornthwaite (1948) and Thornthwaite and Mather (1955). These models are basically bookkeeping procedures that use Equation (8.1) to account for the movement of water from the time it enters a basin as precipitation, to the time it leaves the basin as runoff or ET. The models vary in their degree of complexity depending on the detail with which each component of Equation (8.1) is considered. Most models account for direct runoff from rainfall and lagged runoff from basin storage in the computation of total runoff (Q). In addition, most models compute the ET term as some function of potential evapotranspiration and the water available in storage (S). Although water balance models can be applied at daily, weekly, monthly, or annual time steps, in climate studies they have been applied most frequently at the monthly time step.

A simple three-parameter monthly water balance model was applied by Arnell (1992) to 15 basins in the United Kingdom to estimate changes in monthly river flows and to investigate the factors controlling the effects of climate change on river flows in a humid temperate climate. The three parameters, which were fitted to each individual basin, represent (1) the fraction of precipitation that contributes directly to runoff, (2) the maximum storage capacity of the basin, and (3) the basin lag for converting the water available for runoff to streamflow. Arnell (1992) also compared changes in average annual runoff as computed by the water balance model with four different empirical models for the assumed climate changes. No one empirical formulation gave a consistently closer match to the water balance model estimates of the effects of climate change, and there were large differences among empirical models for the same scenario. According to Arnell, the results suggest that 'estimates of possible change based on annual empirical models should be treated with extreme caution'.

Gleick (1987a) developed a monthly water balance model for application to the Sacramento River basin in California. He began with a basic model similar to that used by Arnell (1992), but varied the parameter representing the fraction of precipitation that contributes to direct runoff by season. A snow accumulation and melt component was developed to account for the seasonal storage and release of water by a snowpack. The model was applied to evaluate changes in runoff and soil moisture using 18 different climate change scenarios (Gleick, 1987b). A monthly water balance model that also accounts for snow processes was developed and applied to the central mountains of Greece by Mimikou *et al.* (1991) to evaluate the regional hydrological effects of climate change.

Schaake (1990) developed a nonlinear monthly water balance model for the evaluation of regional changes in annual runoff associated with assumed changes in climate. The model was applied to 52 basins in the southeastern United States using a single set of model parameters for all basins. The storage term in Equation (8.1) was recast as a deficit term, where the deficit is the difference between a parameter defined as the maximum limit of storage and the current storage. When the deficit is zero, water is assumed to evaporate and transpire at the potential rate, and all precipitation is assumed to be direct runoff. Monthly potential ET was computed from the annual average potential ET using a sinusoidal relation to account for the variation in potential ET over the course of a year. Actual ET was computed as a function of potential ET and the moisture deficit. Runoff was computed as a function of precipitation, the moisture deficit and actual ET.

Water balance models provide the ability to simulate average runoff for given precipitation and temperature over a range of basin conditions, and to simulate the year-to-year variations in runoff as precipitation and temperature vary. The limitations of such models include the need to calibrate parameters to observed conditions, and the inability of monthly or annual time step models to account adequately for possible changes in individual storm runoff characteristics at time steps of a few hours to a few days.

Conceptual lumped-parameter models

Conceptual lumped-parameter models are developed using approximations or simplifications of fundamental physical laws and may include some amount of empiricism. They attempt to account for the linear and nonlinear relations among the components of Equation (8.1). As with water balance models, conceptual lumped-parameter models attempt to account for the movement of water from the time it enters the basin until it leaves as runoff. However, the flow paths and residence times of water are considered in much greater detail and normally at time steps of the order of minutes, hours, or one day. Vertical and lateral flow processes may be considered. Vertical processes may include interception storage and evaporation, infiltration, soil moisture storage, evapotranspiration, groundwater recharge, and snowpack accumulation and melt. Lateral flow processes may include surface runoff, subsurface flow, groundwater flow and streamflow. In addition, some models include the capability to simulate some associated sediment, chemical and biological processes.

Processes are usually parameterized at the scale of an entire basin or for relatively large subareas of the basin. These areas often have heterogeneous mixes of vegetation, soils and land use. Consequently, parameters are assumed to be effective values that are representative of the mix of conditions and normally must be calibrated using historical information. Some applications attempt to account for spatial variability in basin characteristics by making subarea delineations based on considerations such as land use or vegetation. The effect of elevation on climate characteristics such as temperature and precipitation are considered by dividing a basin into elevation bands.

One frequently used model in this group is the Sacramento Soil Moisture Accounting Model (Burnash *et al.*, 1973), which simulates the movement and storage of soil moisture using five conceptual storage zones. The model has 17 parameters that define the capacities and flux rates to and from the storage zones. Němec and Schaake (1982) used the Sacramento model to evaluate the effects of a moderate climate change on the sensitivity of water resource systems in an arid and a humid basin in the United States. Sensitivity was evaluated by simulating the variation in storage–yield relations of hypothetical reservoirs located in each basin.

The Sacramento model has been coupled with the Hydro-17 snow model (Anderson, 1973) by a number of investigators for application to basins dominated by snowmelt. Hydro-17 simulates the accumulation, storage and melt of a snowpack at six-hour time steps using a modified temperature index approach. The hydrological sensitivities of four basins in California were evaluated by Lettenmaier and Gan (1990) using the Sacramento model coupled with Hydro-17. Changes in snowpack water equivalent, runoff, evapotranspiration, soil moisture and flood frequency were examined. The same coupled models were used by Schaake (1990) to evaluate the sensitivity of runoff to climate change in the Animas River basin, a sub-basin of the Colorado. Nash and Gleick (1991) also used the model to evaluate runoff sensitivities in two other headwater basins of the Colorado River. Cooley (1990) applied the model to a headwater basin in southwestern Montana. Panagoulia (1992) used the model to assess the effects of climate change on a basin in central Greece.

Several other models with structures similar to those of the coupled Sacramento and Hydro-17 models, but with different process conceptualizations, have been used to assess the effects of climate change on many world regions. The model of the Institut Royal Météorologique de Belgique (IRMB; Bultot and Dupriez, 1976) has been applied to basins in Belgium (Bultot *et al.*, 1988) and Switzerland (Bultot *et al.*, 1992). The HYDROLOG model (Porter and McMahon, 1971) was applied to two basins in South Australia (Nathan *et al.*, 1988). The HBV model (Bergstrom, 1976) has been applied to basins in Finland (Vehviläinen and Lohvansuu, 1991). The Hydrologic Simulation Program–FORTRAN (HSPF) model (US EPA, 1984) has been applied to a basin in Newfoundland, Canada (Ng and Marsalek, 1992).

The Erosion Productivity Impact Calculator (EPIC) is a model that simulates hydrology, erosion and sedimentation, nutrient cycling, plant growth, tillage, soil temperature and crop management (Williams *et al.*, 1984). EPIC was later modified to enable it to simulate the effects of increased atmospheric CO_2 and climate change on the photosynthetic efficiency of crops, and on water use (Stockle *et al.*, 1992a,b). This modified version was used to investigate the effects of rising CO_2 concentrations and climate change on agricultural productivity in the four-state region of Missouri, Iowa, Nebraska and Kansas (Easterling *et al.*, 1992a,b). Applications were made at the single hectare scale for 49 representative sites throughout the region.

The more detailed process simulation capabilities and higher temporal resolution permitted by conceptual lumped-parameter models enables more detailed assessments to be made of the magnitude and timing of process responses to climate change. For example, at a higher resolution the timing and the form of precipitation (rain or snow) enable the estimation of streamflow timing and of the frequency and magnitude of flood peaks. However, these capabilities are accompanied by an increase in the number of process parameters that must be estimated or fitted, and in the amount and types of data needed to characterize the basins and as inputs to run the simulations.

Process-based distributed-parameter models

These models are firmly based in the understanding of the physics of the processes that control basin responses. Process equations involve one or more spatial coordinates and have the capability to forecast the spatial pattern of hydrological conditions in a basin, as well as basin storage and outflows (Beven, 1985). To facilitate this level of detail in a process simulation, the spatial discretization of a basin may be done using a grid-based approach or a topographically based delineation. In each case, the process parameters are determined for each grid cell or topographic element.

The ability to simulate the spatial patterns of hydrological responses within a basin makes this approach attractive for the development of models that couple hydrological processes with a variety of physically based models of biological and chemical processes. The applicability of models of this type to assess the effects of climate change has been recognized (Beven, 1989; Bathurst and O'Connell, 1992), but so far there have been few applications.

The FOREST-BGC (BioGeochemical Cycles) model is a process level ecosystem model that calculates the cycling of water, carbon and nitrogen through forest ecosystems (Running and Coughlan, 1988). Running and Nemani (1991) used this model to examine climate change induced forest responses for a $1540\,km^2$ region in northwestern Montana using a grid resolution of $1.1\,km^2$. Changes in outflows of water from the soil zone, evapotranspiration and photosynthesis were examined for the region and for individual grid cells.

Major limitations to the application of these models are the lack of availability and poor quality of basin and climate data at the spatial and temporal resolution needed to estimate model parameters and to validate model results at this level of detail. These data requirements may pose a limit to the size of basin in which these models are applied. However, the Système Hydrologique Européen (SHE) model (Abbott et al., 1986) is reported to have been successfully applied to basins ranging in size from $30\,m^2$ to $5000\,km^2$ (Bathurst and O'Connell, 1992).

8.2.2 Hydrological modelling issues

Most of the current modelling approaches have been built upon the use of existing operational models, with modifications as needed to extend their applicability to a wider range of basin conditions and to account for limited knowledge of basin and climate characteristics. Although most reports on the use of such models claim some measure of success and applicability of the methods developed, each also discusses a number of qualifying assumptions and limitations that affect the interpretation of model results. A number of these problems are common to all hydrological modelling applications and are being addressed by a variety of research efforts now being undertaken by the hydrological science community.

The following review of selected topical areas in which problems and limitations have been identified provides an additional measure of the state of the art of hydrological modelling.

Parameter estimation

The variety of modelling approaches used reflects a number of factors, including the objectives of the assessment, data constraints and the spatial and temporal scales of application. Although hydrological models differ in their degree of complexity, they share a common problem. Each has a number of parameters that must be estimated or calibrated for model application. Some parameters are defined as being physically based and are assumed to be measurable from basin and climate characteristics, but many other parameters are less well defined and so are optimized or fitted.

Problems with the use of fitted parameters include the limited length of historical data records, minimal or no information on reasonable values or acceptable ranges of values, the incorporation of model and data errors in parameter values, and the effects of parameter intercorrelations. Intercorrelation can produce compensating errors that inadvertently improve a simulation; for example, underestimation of one parameter may be compensated for by overestimation of another parameter, resulting in the right answer, but for the wrong reason.

Problems of parameter fitting in the use of the Sacramento Soil Moisture Model have been investigated in detail by several investigators. Sorooshian and Gupta (1983) noted that one of the most important problems in the calibration of this type of model is the inability to obtain unique and conceptually realistic parameter sets. The problems were demonstrated to be the result of inadequacies in the structure of the models and the automatic techniques used to calibrate them. In a comparison of seven conceptual rainfall–runoff models, including the Sacramento model, Franchini and Pacciani (1991) found that all the models produced similar and equally valid results in spite of the wide range of structural complexities among the models.

Gan and Burgess (1990) examined the parameterization and model response of the Sacramento model for hypothetical basins. Their findings included that the model was unreliable in predicting hydrological response from extreme rainfall, and that the lumped parameters were dependent on the sequence of climate inputs used in the fitting process. The authors noted that 'this finding should be heeded by modellers who use calibrated conceptual models to explore the hydrological consequences of climate change'.

Although other lumped-parameter models may not have been examined in such detail, similar arguments could probably be made for many of the models in this class, as well as most of the current process-based, distributed-parameter models. The frequent use of these models for climate change assessments, given the noted problems, casts a degree of uncertainty on the results and reflects the lack of more robust hydrological modelling approaches.

In climate change studies, problems with the use of fitted parameters are increased by the fact that the climate and basin characteristics used in the fitting may be different from the climate and basin characteristics that are representative of the period with modified climate. Changes in basin characteristics may be due to both climate and anthropogenic causes, and the ability to distinguish between the effects of each will be important. Criteria to determine the suitability of models for application to the assessment of climate change have been developed by Klemeš (1985). He suggested that the model structure must have a sound physical foundation, that each structural component must permit separate validation, and that the model must be geographically and climatically transferable. Geographic transferability is accomplished through the adjustment of model parameters, and climatic transferability by modifying the input data.

The strong physical foundation of process-based distributed-parameter models provides the capabilities outlined by Klemeš, and suggests that these models would be most appropriate for climate change assessments. In addition, they provide the spatial and temporal resolution needed to couple other water-dependent processes such as biological and chemical processes. With increased model complexity, however, comes the need for an increased number of parameters, more extensive data requirements for parameter estimation and model validation, and a degree of uncertainty given the current state of knowledge of basin processes and process parameterization at this level of detail.

In a critical review of physically based modelling, Beven (1989) argued that while the application of physically based models appeared to be rigorous in principle, there are fundamental problems in their application. These problems include unknowns in the system, over-parameterization of the models, and the implicit lumping of subgrid processes inherent in the numerical approximations used. Grayson et al. (1992) made similar criticisms, and warned that 'in developing and using complex models, there is a danger that computational and conceptual complexity is substituted for accurate representation of reality'.

In the development of models for use in climate change assessments, there is a critical need for process formulations whose parameters can be estimated from measurable basin and climate characteristics. This need exists across the full range of model types. The research tasks outlined by Beven (1989) for distributed-parameter models are equally applicable to lumped-parameter models. These tasks include 'the need for a theory of the lumping of subgrid scale processes, for closer correspondence in scale between model equations and field processes, and for the rigorous assessment of uncertainty in model predictions'.

Scale

A major factor in the development of more physically based models, whether lumped- or distributed-parameter, is the consideration of scale. Interests in the assessment of hydrological impacts range spatially from the local to regional to global scale, and temporally from minutes and hours to days, months, years, and longer. As noted by Klemeš (1983), as one moves from small plots and hillslopes to large basin systems, different sets of physical laws dominate at each scale. Physical laws at larger scales tend to express averages or integrals of those dominant at smaller scales. The relative importance of processes also changes as one moves across scales. For a given model, parameters estimated or fitted for small basins may not be representative of larger basins. Likewise, time is a consideration in parameter estimation, in that the parameters estimated for a daily flow simulation may not be representative for other different simulation intervals. A variety of models and modelling approaches will be needed to address the large number of modelling objectives and scales of application.

Knowledge of physical processes is most extensive at the laboratory, point and small plot scales. Extrapolation of this understanding to larger scales must consider the effects of spatial heterogeneity and the relative importance of processes in the parameterization of processes at the larger scales. Wood et al. (1988, 1990) investigated the modelling of basin runoff, and suggested that at some scale, termed a 'representative elementary area' (REA), knowledge of the pattern of heterogeneity within a spatial element is no longer necessary; only the statistical representation of the factors controlling runoff needs to be quantified. This raises questions regarding the validity of the use of 'effective' parameter values to simulate a process over heterogeneous areas. Research remains to be done to define and test the statistical formulations of spatial heterogeneity.

Model evaluation (validation)

The models used to assess the effects of climate change are currently evaluated on the basis of their ability to reproduce historical time series of observed streamflow or other hydrological variables. However, for conditions that are representative of a potential climate change, observations will not be available *a priori*, and the climate, as well as the physical characteristics of a basin, may be significantly different from those used in the parameter calibration procedure. The problem of defining quantitative measures of model performance in terms of its ability to adequately simulate new conditions is formidable.

Klemeš (1985, 1986) suggested a hierarchical scheme for the systematic testing of climatic and geographic transferability in hydrological simulation models. This scheme presents two tests for application under stationary conditions, and two tests for nonstationary conditions. The nonstationary tests are those designed for testing models developed for climatic and geographic transferability.

The first test for nonstationary conditions is a differential split-sample test and is used to evaluate climate transferability. In this test, two periods with different values of the climate parameter of interest are identified. For example, a wet and a dry period are identified if precipitation change is being considered. If the model is to be used to simulate a wetter climate, it should be calibrated using the dry period and validated using the wet period. For an intended model application to a dry period, the procedure is reversed.

The second test is intended to evaluate both climatic and geographic transferability, and is a proxy-basin differential split-sample test. Two basins within a region are selected and, as in the previous test, two periods with different values of the climate parameter of interest are identified. Using the wet and dry example, the model would be calibrated on the wet period of each basin and evaluated on the dry period of the opposite basin. Calibration on the dry period and testing on the wet period using the same paired basin scheme provides an alternative application of this test. Acceptance in the nonstationary tests is based on model performance using the alternative climate and basin conditions.

Determination of the acceptability of test results is ultimately a subjective decision on the modeller's part, based on some criteria he or she establishes. Klemeš (1986) explicitly points out that the testing scheme is intended for models whose outputs are for 'use outside hydrology', which he defines as applications for planning and operational decisions. These tests are considered a minimum requirement for the evaluation of a model's simulation capabilities. In addition, the testing procedures measure only the correctness of estimates of hydrological variables, and not the structural adequacy of the model.

A procedure for calibrating and uncertainty estimation of distributed-parameter models based on generalized likelihood measures was recently developed by Beven and Binley (1992). This approach, termed the Generalized Likelihood Uncertainty Estimation (GLUE) procedure, provides a methodology for evaluating the uncertainty limits of the model for future events for which observed data are not available. The appeal of this approach is that one can calculate quantitative measures of uncertainty that could be used in applying simulation results to water resources management questions. While this method addresses the problem of climate transferability, research remains to be done to evaluate the uncertainty for changes in basin parameterization.

Data

A major limitation to process conceptualization and parameter estimation is the lack of sufficient data to improve our understanding of the fundamental hydrological processes across a range of spatial and temporal scales. In the research and development of improved physically based models, numerous researchers have noted the urgent need for the establishment of field-based data collection programmes jointly with model development efforts (e.g. Dunne, 1983; Beven, 1989; Dozier, 1992; Grayson *et al.*, 1992).

To address the need for field-based data, one approach has been the development and conduct of coordinated large-scale field experiments. Several such studies have been established. The International Satellite Land Surface Climatology Project (ISLSCP) was organized to monitor the variables that govern climate and climate fluctuations at a range of scales from point and plot to regional and global. The First ISLSCP Field Experiment (FIFE) conducted at a tall-grass prairie site in Kansas investigated water and energy flux processes from a scale of cm at the individual plant level to tens of km for a regional perspective. The Boreal Ecosystem–Atmosphere Study (BOREAS) is a similar type of study of two boreal forest regions of Canada that is an interdisciplinary investigation of land surface climatology, tropospheric chemistry and terrestrial ecology.

The Global Energy and Water Cycle Experiment (GEWEX) Continental-Scale International Project (GCIP) is using the Mississippi River basin as a continental-scale area in which to conduct a variety of coordinated studies. The objectives of these studies include: to determine the time–space variability of hydrological and energy budgets over regional and continental scales; to develop and validate macro-scale hydrological models and coupled hydrological/atmospheric models; and to provide a capability to translate the effects of possible future climate change into impacts on water resources on a regional basis (WCRP, 1992).

One area in which technical advances in data collection will have a significant effect on hydrological modelling is remote sensing. Remote sensing offers the possibility of obtaining frequent hydrological measurements over wide spatial scales. Data include the presence or absence of vegetation, vegetation structure, vegetation moisture stress, soil type, snow cover, snowpack water equivalents, and soil moisture (Engman and Gurney, 1991; Dozier, 1992). Quantitative precipitation estimates at a spatial resolution of about 4 km and a temporal resolution of 5–6 minutes will be generated by the next generation of weather radar (NEXRAD), and will be used to produce hourly, three-hourly, and storm total products (Alberty et al., 1991).

Programmes such as FIFE, BOREAS and GCIP, with their extensive databases of simultaneous observations from ground stations, atmospheric soundings, aircraft and satellites, will be important sources of the types of information that are needed to address the issues of parameterization and scaling that are so critical to the improvement of the understanding and modelling of hydrological, climatic and biogeochemical processes. To maximize the benefits of data from these and other investigations, improved information management systems are needed to enable the integration of data collected at different scales and by different agencies, and to make these data available to the scientific community on a timely basis (Dozier, 1992).

8.3 ATMOSPHERIC MODELS

The atmospheric models available for modelling climate range from one-dimensional radiative–convective and energy balance type models, to full three-dimensional GCMs (Henderson-Sellers and Robinson, 1992). GCMs solve the equations for the conservation of mass, momentum and energy at the Earth's surface and at a number of heights in the atmosphere. They are able to consider the radiative effects of atmospheric gases and clouds and the interactions of the atmosphere with the land surface, oceans, sea ice and snow cover.

The ability of GCMs to consider explicitly changes in atmospheric composition of radiatively active gases such as water vapour, carbon dioxide and ozone, is a major strength cited in support of the use of GCMs to develop climate change scenarios. However, a number of weaknesses in current GCMs, including their coarse spatial and temporal resolutions, and inconsistencies among different models as to the magnitude and direction of climate change at both regional and global scales, have limited the use of their outputs to little more than estimates of the magnitude of the problem. This section provides a brief review of GCM modelling approaches and the limitations to their current use.

8.3.1 Approaches

Atmospheric circulation at the global scale is driven by a number of large-scale forcings that are either external or internal in origin. Solar radiation is the primary external forcing, while energy, water and gas fluxes between the atmosphere and ocean, land, snow and ice surfaces provide the internal forcings. Radiatively active gases such as CO_2 regulate the amount of solar and terrestrial radiation that is absorbed by the atmosphere, and thus play a major role in the dynamics of the atmosphere. Three-dimensional atmospheric GCMs are used to simulate these processes and their interactions. Current GCMs represent the Earth's surface at horizontal resolutions ranging from 2.5 to 8° and represent the vertical structure of the atmosphere with from 2 to 30 layers (Gates, 1995b).

The differences among GCMs are often a function of the differences in conceptualization and parameterization of the atmospheric physics and surface–atmosphere exchange processes being simulated. At their current scale of horizontal resolution, GCMs simulate the effects of large-scale forcings on atmospheric circulation but miss a number of local forcings that act at the scale of a few to tens of kilometres, and which strongly affect climate variables at the regional scale (Giorgi and Mearns, 1991). These local forcings are induced by the distributions of surface characteristics such as topography, coastlines, water bodies, vegetation and urban areas. Within GCMs these small-scale processes are typically ignored, coarsely discretized, or represented by empirical or statistical relationships developed from observed data. The level of sophistication of these parameterizations varies among models as well. The major processes and process interactions of concern to climate modellers, and the basic structure of GCMs, have been reviewed by a number of authors (e.g. Verstraete and Dickinson, 1986; Dickinson, 1988; Mitchell, 1989; Avissar and Verstraete, 1990; Crane, 1992).

Although some of the small-scale features that affect local forcings are natural, others are strongly affected by anthropogenic changes in surface land use (Avissar and Pielke, 1989). There is an urgent need to develop modelling capabilities to distinguish between and simulate the effects of natural and anthropogenic influences on climate, given the large-scale changes in land use that have occurred and continue to occur around the world.

One approach that addresses the need for smaller-scale resolution in regional analyses uses a GCM with a variable horizontal resolution. Here the resolution is gradually increased towards a point in a region of interest, while on the opposite side of the Earth the resolution is correspondingly decreased so that the mean resolution is unchanged (Courtier and Geleyn, 1988).

A second approach, referred to as 'nesting', is to embed a higher-resolution local area model (LAM) of a selected region in a lower-resolution GCM. Information from the GCM is used to drive the LAM by establishing the initial conditions and providing the boundary conditions for each time step. Information may move interactively between the two models in a two-way coupling, whereas in one-way coupled systems the information flow is only from the GCM to the LAM.

The NCAR nested modelling system is a one-way nested approach that couples a GCM (the NCAR Community Climate Model, version CCM1; Williamson et al., 1987) with a LAM, the mesoscale model MM4 (Anthes et al., 1987). Significant improvements in the simulations of average monthly values of regional climatic detail for the western United States and Europe were demonstrated using the NCAR nested system as compared with the results from the GCM alone (Giorgi and Mearns, 1991). The grid spacings used in MM4 were 60 km for the western United States and 70 km for Europe. It was also suggested that, with the addition of extra physics modules for ecosystem and surface hydrology models to MM4, this approach could be used to investigate the effects of climate change on local environments, and possibly to incorporate feedback mechanisms from these processes into MM4.

A similar nested model approach is being developed for Europe in which a high-resolution LAM called HIRHAM is driven with initial fields and boundary values from a GCM (Machenbauer et al., 1996). In current tests the grid spacing for HIRHAM is about 50 km.

The effects of increasing CO_2 concentrations on climate are assessed by first running a GCM for a multi-year period using current CO_2 concentrations. The model equations are initialized in space and integrated forward in time until they become independent of the initial conditions and reach an equilibrium that is representative of the mean climate for the given set of boundary conditions. Time-averaged values, as well as statistical measures of short-term variability, of circulation, temperature and other variables of interest are then computed on an annual, seasonal and/or monthly basis using all the years in a long-term simulation. The concentration of CO_2 in the modelled atmosphere is then increased to some anticipated level and the GCM is run again for the same period, producing equivalent time-averaged and short-term variability values. Comparing these values between runs provides some measure of the effects of the prescribed increase in CO_2.

A number of studies in the mid-1980s simulated the effects of a doubling of CO_2 ($2 \times CO_2$) on the mean global changes in surface temperatures and precipitation. The models used included the GCMs of the Goddard Institute for Space Studies (GISS; Hansen et al., 1984), the National Center for Atmospheric Research (NCAR; Washington and Meehl, 1984), the Geophysical Fluids Dynamic Laboratory (GFDL; Weatherald and Manabe, 1986), the UK Meteorological Office (UKMO; Wilson and Mitchell, 1987), and Oregon State University (OSU; Schlesinger and Zhao, 1987). The results showed temperature increases ranging from 2.8 to 5.2°C, and precipitation increases ranging from 7.1 to 15%. Although the large-scale features of temperature were consistent among the models, there were wide regional differences among the models (Mitchell, 1989).

The $2 \times CO_2$ experiments reflect the possible effects on climate at some distant time in the future. However, an increase in CO_2 to twice the current level will not occur as a single step function, but rather will occur gradually over time. Transient response experiments have also been conducted in GCMs, in which CO_2 is increased in increments over time and the resulting changes in climate are evaluated (Washington and Meehl, 1989; Stouffer et al., 1989).

As noted by Mitchell (1989), trust in the results of GCMs rests on the extent to which they are based on physical principles and their ability to reproduce the present climate, including its temporal and spatial variations. A number of studies comparing GCM outputs with observed climate data have been conducted over the past few decades. These studies raise a number of issues regarding the accuracy and applicability of GCMs to regional and global assessments of the effects of climate change on hydrology.

8.3.2 Atmospheric modelling issues

Simulations of current climate

Comparisons of GCM outputs with observed climate data have raised a number of issues regarding the use of their outputs in climate change impact assessments. For example, a comparison of the average seasonal surface air temperatures computed by four GCMs indicated that while there was general agreement among the models and with historical data on surface air temperatures at large scales, there were significant differences at the regional to local scales (Grotch, 1988). The magnitudes of these differences were such that serious questions were raised as to the applicability of model results at the regional and finer scales that are usually of interest in hydrological analyses.

Such comparisons also provide valuable insight into model performance and provide the basis for model improvements over time. To provide a systematic and comprehensive intercomparison of atmospheric climate models, the Atmospheric Model Intercomparison Project (AMIP) was organized by the Working Group on Numerical Experimentation, as a contribution to the World Climate Research Programme (Gates, 1992). AMIP is an international effort to investigate the systematic climate errors of GCMs on seasonal and annual time scales. Some 30 models were included in the study, each simulating the climate of the period 1979–88 using a common set of observed monthly averaged distributions of sea surface temperatures and extent of sea ice as boundary conditions. A common set of standard monthly averaged output was defined for the project and all model results were analysed using a common set of diagnostic measures of performance.

Within the AMIP a number of diagnostic subprojects were set up to perform detailed examinations of model performance and simulations of specific physical processes and process interactions. Summaries of model performance and of the detailed analyses of comparative model behaviour by these subprojects are presented in the *Proceedings of The First International AMIP Scientific Conference* (Gates, 1995a). A review of these findings indicates that although some improvements in GCM performance have been made over time, a number of problems remain to be resolved before simulations of global and regional climates can be regarded as adequate with any degree of confidence.

The interactions of the atmosphere with the hydrosphere, cryosphere and land surface are key areas in which our understanding of the processes and process interactions involved needs to be improved in order to develop improved GCMs. In order to provide a general sense of the state-of-the-art of GCM climate simulations, a brief summary of some of the initial findings of the AMIP in these areas is presented in the following.

Hydrosphere

The oceans are a large component of the hydrosphere and play a major role as sources and sinks and in the transport of water and energy in the global system. It was recently shown that the global temperature record for the period 1979–92 could be reproduced by an atmospheric model forced only with ocean surface temperatures (Graham, 1995). Such a finding demonstrates the importance of the oceans in the global system, and supports the need for a fully coupled ocean–atmosphere model to adequately simulate the global climate.

In AMIP, simulations of ocean surface heat fluxes and the implied meridional heat transport between the atmosphere and ocean were evaluated by Randall and Gleckler (1995). Ocean general circulation models (OGCMs) have been developed to simulate the ocean energy flux and transport processes. However, in applications of coupled atmospheric and ocean GCMs with no boundary constraints, the simulated climate typically drifts toward an unrealistic state. To prevent this drift most coupled models apply some form of flux correction to the computations. The AMIP programme used an uncoupled approach by prescribing sea surface temperatures (SSTs) and sea ice distributions.

The AMIP analyses showed that with these prescribed SSTs, the GCMs produced surface energy budgets that implied ocean energy fluxes that varied widely among models, and these differences were related to model-to-model differences in cloud radiative forcing. It was suggested that more realistic cloud radiative forcings in future models should improve the simulated surface energy budgets and coupled model simulations without the need for flux corrections.

Cryosphere

The Arctic and Antarctic regions are thermal sinks for the northern and southern hemispheres. Changes in the climates of these regions would modify the thermal gradients between the equator and poles and could change climate over the mid-latitudes. Changes in temperature and precipitation could also modify the thermohaline circulation in the oceans due to changes in the freshwater inflows to the polar oceans, and the areal extent and thickness of sea ice (WMO, 1992b).

Analyses of the results for the Arctic region showed that there was large model-to-model scatter in the simulated variables and major differences between simulated and observed variables such as surface air temperature, precipitation and cloud cover (Walsh *et al.*, 1995). The range of simulated zonal mean surface air temperatures among the models exceeded 10°C at nearly all latitudes and for nearly all seasons. Simulated cloud cover showed large variations in magnitude, and the observed seasonality in cloud cover was demonstrated only qualitatively in some models, while others showed a seasonality opposite to that observed. Precipitation was overestimated in all the models, especially during the winter months. The difference between precipitation and evaporation was about twice the best estimates from observations, which would indicate that simulated freshwater inflows to the Arctic Ocean would also be overestimated.

Hydrological and land surface processes
The ability of GCMs to simulate the global hydrological cycle and the associated variables temperature, precipitation and evaporation was investigated by Lau *et al.* (1995). The differences among the models included a range of global mean surface air temperatures of 6–7°C and of up to 2 mm in precipitation. In general, the models had a cold–wet bias. The global mean precipitation over the oceans was greater than over land for most models, in agreement with observations, but over land, most models produced more precipitation than observed and this was generally accompanied by excessive evaporation. In addition, there was a systematic underestimation of the number of storms with a rainfall rate of 1 mm or less, which suggested some fundamental deficiencies in the rainfall-related processes of the models. It was noted, however, that the uncertainties in the observed global surface air temperatures and precipitation are also quite large.

There were also large variations among the models and in comparison with observations in the global distributions of annual and seasonal precipitation. The areas of largest disagreement among the models and with observations were in the eastern tropical Pacific intertropical convergence zone, the southern Pacific convergence zone and the Asian monsoon region. The annual harmonics of the precipitation cycle were reasonably well simulated in regions where there are strong annual cycles or monsoon characteristics, but were more divergent in areas where strong cycles are absent. Major discrepancies were noted in the Mississippi basin and the Tropical Oceans Global Atmosphere

(TOGA)-Coupled Ocean–Atmosphere Experiment (COARE) study regions.

The spatial and interannual variability of simulated tropical precipitation was compared with land-based observed climatology by Srinivasan *et al.* (1995). These regions also showed large inter-model variations, with some models doing a reasonable job of capturing mean spatial patterns of tropical precipitation. There was no overall relationship between model resolution and model performance, and only three of the models compared were able to reproduce the interannual variability of seasonal precipitation in the central Pacific region where El Niño is known to be a factor.

The amount and the distribution of water vapour in the atmosphere play a major role in precipitation, evaporation, and other water and energy flux processes. Simulated monthly mean gridded precipitable water, zonal mean specific humidity and zonal mean meridional water vapour fluxes were compared with three sets of observed data for different regions by Gaffen *et al.* (1995). The results showed that the models tended to underestimate the precipitable water over North America, globally, and in the zonal band 35–50°N. Both models and observations showed ENSO signals and trends in precipitable water, but the models did not capture much of the higher-frequency interannual variations. In addition, the models appear to overestimate the poleward flux of moisture, which is consistent with the overestimate of Arctic precipitation discussed above. It was suggested that model differences may be linked to differences in the treatment of boundary layer processes and in the treatment of soil moisture.

The simulated spatial patterns, mean annual cycles and interannual variations in soil moisture were compared with observed soil moisture data sets from Russia, China and the state of Illinois, USA (Roback *et al.*, 1995). The AMIP survey identified seven different classifications of soil moisture schemes among the GCMs. The results of all of the models were quite different from observations, and there was no evidence that more complicated schemes produced better simulations than those based on the basic 'bucket' model approach of Manabe (1969). A value of 15 cm, which is used in many of the bucket models as a global value for field capacity, was found to be too low to capture the variations in soil moisture at high latitudes. In addition, the observed variations in soil moisture in winter at these latitudes were not properly simulated by any of the models. The amplitude of the seasonal cycle of soil moisture did not compare with observations at any of the data sites, but the general phase of the cycle was correct. Interannual variations in soil moisture for all three data sets were also not captured by any of the models.

PILPS

Whereas the AMIP is focused on comparisons of GCMs, a related project, the Project for Intercomparison of Land Surface Parameterization Schemes (PILPS) is conducting a comparison of land surface schemes with the purpose of evaluating and improving these schemes for climate and weather prediction models (Henderson-Sellers *et al.*, 1993; 1995). In phase 1 of PILPS, 26 models were run 'off-line', i.e. they were not coupled with an atmospheric model. In simulations of three different environments (tropical forest, mid-latitude grassland and tundra) large differences were found among the models in flux rates and other variables. In phase 2 the models are being compared using observation data from selected field sites where these data are available for both model input and evaluation.

Phase 3 of PILPS is associated with the AMIP in that this phase is an intercomparison of 11 land surface schemes that are coupled with their host 3D GCMs participating in AMIP. Analyses of overall land surface climatology in the AMIP simulations were used to identify regions for further study within PILPS (Love and Henderson-Sellers, 1994). Consistencies and differences in the simulated values for net radiation, precipitation and sensible and latent heat fluxes were used in the selection process.

A comparison of the AMIP results for the three areas used in phase 1 of PILPS showed that (1) model differences in the simulated surface temperatures are generally larger when the land surface schemes are interactive components of GCMs, and (2) more heat was lost from the land surface in the form of sensible heat in the coupled AMIP study than in PILPS phase 1 (Henderson-Sellers *et al.*, 1995). It was also suggested that differences among the AMIP GCM results may have been exaggerated by the differences in land surface parameterization schemes. Phase 4 will investigate the coupling of selected PILPS schemes with selected GCMs and will enable a more detailed analysis of the effects of the different schemes on GCM results.

AMIP and PILPS are complementary efforts to evaluate and improve atmospheric modelling capabilities. Whereas AMIP addresses the accuracy of larger-scale GCM outputs, PILPS phases 1 and 2 address land surface parameterizations at a smaller scale. Joint AMIP–PILPS efforts are focused on the scaling up and incorporation of selected land surface parameterization schemes into GCMs in order to identify those schemes that are most appropriate for improving GCM performance. It is only through efforts such as these to evaluate model performance, with field observations at a variety of temporal and spatial scales, that improvements in current modelling methods can be made, and that confidence can be established in the use of model results for impact assessments.

Such confidence from the hydrological community will require that the validity of atmospheric model output be demonstrated at a wide range of temporal and spatial scales. In addition to the comparison of average annual and seasonal values, it needs to be demonstrated that reliable output time series of climate variables that are typically used as inputs to hydrological models can be produced at time steps of daily or shorter time increments. Such time series must have the appropriate inter- and intra-annual variability, and the appropriate magnitudes and frequencies of climate extremes that produce events such as floods and droughts. One such demonstration would include the simulation of historical El Niño and La Niña sequences and their associated regional effects on temperature and precipitation.

8.4 CLIMATE SCENARIO DEVELOPMENT

Climate scenarios are sets of time series or statistical measures of climate variables, such as temperature and precipitation, that represent alternative climate conditions for past, present or future conditions. A variety of methods have been developed to generate climate scenarios for use in the assessment of the impacts of climate variability and change. These methods are typically categorized as being empirical or process model based. The empirical approaches use information about the past to estimate possible future conditions, whereas model-based approaches attempt to use the underlying physical laws of atmospheric processes to simulate climate under changed conditions. These approaches have been reviewed by Robinson and Finkelstein (1989) and Carter *et al.* (1994).

8.4.1 Empirical methods

Compared to model methods using GCMs, empirical methods provide a greater spatial and temporal resolution for many regions but cannot directly consider changes in atmospheric composition, and specific processes are deduced only by inference (Robinson and Finkelstein, 1989). One empirical approach is simply to make 'educated' estimates of potential changes in air temperature and/or precipitation and then to use these estimates to adjust the available historical record. These hypothetical scenarios are typically used to evaluate the sensitivity of a region or a selected assessment model by selecting and applying a range of estimated values.

Another approach is the statistical synthesis of scenarios. Statistical scenarios preserve historically observed statistics while providing the ability to develop time series that are longer than the observed record. New sequences of events can also be developed by changing the statistical parameters of the historical record. However, synthetic time series fail to capture all of the complexity and correlations that exist in the historical record.

Analogue techniques are an additional empirical approach. Temporal analogues use records from the past as an estimate of future climate, whereas spatial analogues use current records from another region as an estimate of future climate in the given region. Problems with these approaches arise from the fact that for temporal analogues, the physical mechanisms and boundary conditions that gave rise to the historic conditions may have been quite different from those of today; for spatial analogues important factors such as the terrain, day length and economic development of the regions may differ substantially (Carter *et al.*, 1994).

8.4.2 Process models

Process models allow the explicit incorporation of many atmospheric processes and changes in atmospheric composition, but are limited by their coarse temporal and spatial resolution and the large degree of uncertainty in their results. Although most GCMs simulate reasonable average annual and seasonal features of present climate over large geographical areas, they are less reliable in simulating smaller spatial and temporal scale features that are relevant for impact assessments (Grotch and MacCracken, 1991). In addition, the outputs from different GCMs can vary significantly for some regions, posing the problem of which GCM can be regarded as correct.

Given these constraints, one of the most widely used methods of scenario generation has been to estimate average annual changes in precipitation and temperature for a region using one or more GCMs, and then to apply these estimates to adjust historical time series of precipitation and temperature. In the simplest procedure, the adjustment is made by multiplying the historical precipitation by a percentage change and adding an absolute change to the historical temperature.

These procedures account for changes in the mean of the historical time series, but do not provide for changes in the variance. The frequency of extreme events is relatively more dependent on changes in the variability than in the mean of climate (Katz and Brown, 1992). In turn, the resulting changes in the frequency and magnitude of storms and

droughts may have a greater impact on water resources than those resulting from only a change in the mean. Changes in the variance have been implemented in some hydrological assessments by varying the magnitudes of temperature and precipitation changes on a monthly basis (Gleick, 1987a; Bultot *et al.*, 1988; Arnell, 1992).

To address the variability problem, a number of methods have been developed to disaggregate the GCM output for direct application in hydrological models. Hay *et al.* (1992) developed a methodology to disaggregate GCM precipitation over the eastern United States using six weather types, classified on the basis of wind direction and cloud cover, and the precipitation characteristics of each weather type. Variations in weather type frequencies and characteristics are determined from GCM outputs for current and future conditions and are used in a stochastic precipitation model to predict current and future precipitation. A conceptually similar approach was developed for the western United States by Dettinger and Cayan (1995).

Wilson *et al.* (1992) developed a stochastic model of weather states and concurrent daily precipitation at multiple precipitation stations. Weather states were computed using grid cells and grid variables from the US National Meteorological Center grid point data set. Proposed applications of this procedure include the use of grid variables from GCMs to investigate the implications of alternative climates to both the weather states and to station precipitation.

Wilks (1992) presented a method to adapt stochastic weather generation models to generate synthetic daily time series of temperature, precipitation and solar radiation, consistent with assumed future climates as modelled by GCMs. The parameters defining the current daily stochastic process at a site are adjusted based on monthly GCM values to produce changed climate scenarios. The model is then run in a Monte Carlo sense to produce streams of daily weather values for periods of arbitrary length. Katz (1996) showed the importance of having a complete understanding of the statistical properties of the selected weather generator. The theoretical statistical properties of one selected weather generation model were derived, and it was shown that when its parameters were varied, certain unanticipated effects can be produced. In this case, modification of the probability of the occurrence of precipitation changed the mean, variance and autocorrelation of daily temperature. Using a probabilistic model for the western United States, Dettinger and Cayan (1992) found that such changes were a natural outgrowth of the underlying changes in atmospheric circulation patterns.

Climate scenario estimates developed from GCM results are based on an assumption of the accuracy of the GCM which may or may not be warranted. Improved climate scenarios will be a function of both improved disaggregation techniques and improved simulation capabilities of the GCM. In addition, future scenario estimates must include not only reliable estimates of temperature and precipitation, but should also include the associated values of humidity, wind, and short- and long-wavelength radiation. These values will enable the development and application of more robust hydrological models for the assessment of a far broader range of climate change-related impacts.

Development and testing of scenario estimation methodologies in all the major climatic and physiographic regions are needed to assess fully their capabilities, to define the more robust approaches, and to improve the resulting estimates of climate change at the range of scales required by hydrological models. The results of such tests are also needed to define and establish a more consistent set of approaches for climate scenario generation in all world regions to facilitate inter- and intra-regional comparisons of the impacts of climate change.

8.5 INTEGRATION OF HYDROLOGICAL AND ATMOSPHERIC MODELS FOR IMPACT ASSESSMENT

Given the uncertainties in GCM performance, their use in hydrological impact assessments has generally been limited to providing measures of the magnitude and direction of changes in mean annual and/or seasonal temperatures and precipitation. However, an alternative approach is to use integrated atmospheric and hydrological models directly in the assessment process. Integrated systems range from running the models sequentially as a simple, loosely coupled system, to running them together as a complex, fully coupled system that includes feedbacks between models.

Large-scale GCMs have hydrological components and some have been used to investigate river runoff for current and $2 \times CO_2$ conditions for the largest rivers of the world (Miller and Russell, 1992; Kuhl and Miller, 1992; Van Blarcum et al., 1995). However, the spatial and temporal scales of GCMs are too coarse to provide reliable assessments for smaller-scale basins which are the concerns of resource managers and users. At the same time, the hydrological models used in assessment studies operate at scales that are too small for their direct incorporation into GCMs. As discussed by Hostetler (1994), the problem of discordant scales between atmospheric and hydrological models is being addressed by scaling up or aggregating hydrological models and scaling down or disaggregating GCMs.

A primary method of disaggregating GCM output is through the use of the nested modelling approach described in Section 8.3 in relation to atmospheric models. Current nested model studies have focused on the performance of the LAMs embedded within GCMs. In these studies the LAMs are driven with observed boundary conditions to eliminate possible sources of error from the coupled GCM, and to enable a comparison of the LAM results with observed meteorological data from the region being simulated.

The feasibility of coupling the output from the MM4 LAM with catchment-scale hydrological models was explored by Hostetler and Giorgi (1993). Simulated daily average values of air temperature, humidity, wind speed and short- and long-wave radiation were used to drive a lake model for Lake Pyramid in Nevada. Monthly average values of simulated solar radiation and precipitation were used to drive a monthly streamflow model for a $600 \, km^2$ watershed in Oregon. Annual totals of lake evaporation simulated using MM4 output agreed well with those simulated using observed data. However, winter precipitation and the range of variability in monthly precipitation were underestimated at the Oregon site, resulting in simulated streamflow being 20% less than observed. Topographic smoothing at the 60 km resolution of the MM4 application, was noted as a possible cause of reduced precipitation due to the resulting reduced effects on topographic forcing.

Topographic effects occur at scales of a few kilometres, but play an important role in determining the magnitude and distribution of precipitation in mountainous terrain. One approach to considering these smaller-scale processes was to include a third model in the nested model approach (Leavesley et al., 1992). An orographic–precipitation model was added to the nest and was driven by output from the LAM. The orographic–precipitation model output, at a resolution of 5 or 10 km, was then used as input to a distributed-parameter hydrological model. The results of a simulation of streamflow using the modelled precipitation values were similar in terms of volume error to those simulated using measured point precipitation. However, the improved information on precipitation distribution provided by the precipitation model resulted in significant improvements in the simulation of the magnitude and timing of streamflow.

Leung et al. (1996) addressed the problem of orography by developing a subgrid orographic–precipitation model which was then included in the Pacific Northwest Laboratory (PNL) regional climate model. The scheme partitions a grid cell into elevation classes, and radiative transfer, turbulent mixing, convection and land surface physics are computed for each elevation class. A full set of hydrological variables, including precipitation, snow water equivalent, soil moisture and surface runoff, are simulated for each elevation class as well. The outputs can also be used as inputs to the PNL's DHSVM watershed model for watershed-scale evaluations.

The process of aggregating small-scale hydrological process model formulations to macroscale formulations suitable for incorporation in a GCM has been a focus of a number of climate research programmes such as FIFE, BOREAS and GCIP. A research product of FIFE was a multiscale model that uses a topographic framework to aggregate a soil–vegetation–atmosphere transfer scheme from the local to the catchment scale, and a statistical–dynamical approach to aggregate to the macroscale (Famiglietti and Wood, 1994a,b). Research is also being conducted independently of these large projects, and the extent of this research is demonstrated in the large number of land surface models participating within PILPS. References to these models can be found available in Henderson-Sellers et al. (1995).

While the atmospheric and hydrological research communities continue to collaborate on model improvement, a fully coupled integrated atmospheric and hydrological modelling system remains an elusive goal. As noted by Hostetler (1994), hydrological models explicitly represent small-scale processes and parameterize larger-scale ones, whereas atmospheric models explicitly represent large-scale processes and parameterize small-scale ones. Thus, when coupling the two modelling approaches, the problem of scales is most pronounced at the juxtaposition of the most highly parameterized processes in both models. Such a coupling has the potential to maximize the errors associated with these parameterizations in each model.

One approach to the scaling issue is to use finer scales of resolution within the GCM to make them more compatible with the hydrological resolution. However, this raises a number of issues, including the potential need for changes in process formulations as the horizontal scale decreases, and the availability of sufficient computing power to handle the computational load at these finer resolutions. The nested modelling approach appears to provide a reasonable alternative to providing finer-scale resolution in the region of interest, but the results are currently limited by the quality of the GCM outputs which are providing the boundary conditions for the regional model.

A large number of problems remain to be resolved in the development of integrated modelling systems for impact assessments. However, the magnitude of these problems has provided a focus for both the atmospheric and hydrological science communities, and has facilitated the organization and conduct of major interdisciplinary studies to address these problems.

8.6 SUMMARY AND CONCLUSIONS

Numerous atmospheric and hydrological models and modelling approaches are currently being used to assess the impacts of climate change on water resources. Information on which model and/or modelling approach to select for specific assessment needs, however, is sorely lacking.

Hydrological model selection has been generally limited to the use of operational models that have been tested in a wide variety of geographic regions or extensively in the region of interest. Model choice is normally a function of problem objectives, data constraints and the spatial and temporal scales of application. Empirical and water balance type models have generally been applied to large-basin and regional analyses at time scales of months to seasons to years. More detailed conceptual lumped-parameter and process-based distributed-parameter models have generally been applied to smaller basins at time scales of 24 hours or less.

Empirical models have minimal data requirements but have questionable transfer value to basin and climate conditions different from those used to develop the input–output relations of the model. An increased range of simulation capabilities is provided as one moves from water balance models to conceptual lumped-parameter models to process-based distributed-parameter models. This is accompanied, however, by an increase in the number of model parameters and in the amount and types of data needed to support parameter estimation and model operation. Limited knowledge of the relations between parameter values and measurable basin and climate characteristics often results in many of the parameters being calibrated to measured data. The use of calibrated parameters also provides a large degree of uncertainty in the ability of these types of models to be climatically and geographically transferable.

A systematic analysis of hydrological modelling capabilities is needed to identify which models and modelling approaches are most robust for the wide variety of impact assessment studies being conducted. Such assessments should not be limited to current operational models, but should include models that can take advantage of the additional climate variables such as humidity, wind and radiation; although these are not widely available in observation data sets, they can be obtained from atmospheric models. Such an analysis would be a valuable tool to identify research needs and to advance the science of hydrological modelling. A PILPS-type approach to investigate hydrological models would provide the focus for such an effort and would complement much of the ongoing research in climate change.

The selection of an atmospheric model for impact assessment is difficult, given the wide range of model results and the large degree of uncertainty in the results. The inability to simulate adequately the current climate, the large differences among models in their predictions of the effects of increasing greenhouse gas concentrations at global and regional scales, and the inability to provide output that is spatially and temporally compatible with the full range of hydrological modelling needs, are major limitations to the use of atmospheric models in impact assessment studies. Some of these problems are being addressed through model intercomparison studies such as AMIP and PILPS. These studies provide valuable insights into processes and process interactions, and guide model improvement by identifying model errors and additional research needs. However, field studies supporting these efforts are limited in the number and types of areas studied and in their duration.

GCMs exhibit their greatest signal-to-noise ratio at the global scale, with a decrease in this ratio as one moves to the GCM grid cell or smaller scale (Ghan, 1992). This is a critical issue for assessments of the hydrological impacts of climate change, since typical hydrological applications are at a range of smaller temporal and spatial scales. Higher-resolution GCMs and nested modelling approaches are being developed to address the problems of discordant scales between atmospheric and hydrological models. At the same time, hydrological models are being scaled up to provide macro-scale hydrology models that are compatible with GCM scales. A large number of interdisciplinary research efforts are now in progress to improve models for use in impact assessments. However, much still remains to be done to improve the signal and to minimize the noise in the output of atmospheric models. Until such improvements are made, a large degree of uncertainty will remain in the results of impact assessments that are made using GCMs.

A review of current atmospheric and hydrological modelling studies indicates a number of problem areas common to the variety of models applied. To address these problems, further research is needed:

- To develop a more physically based understanding of atmospheric and hydrological processes and their interactions. The complexities of atmospheric and hydrological systems are such that process parameterizations will always represent an integration of the spatial heterogeneity of the factors that control these processes. However, when these parameterizations are based on the physics of the process, the ability to measure or estimate parameter values from climate and basin characteristics is improved. It is only through the use of parameterizations that do not require calibration that the problems of climatic and geographic transferability will be resolved.

- Parameter measurement and estimation techniques must be developed for applications over a range of temporal and spatial scales. In moving from point to hillslope to basin to GCM grid cell, and from minutes to centuries, different sets of physical laws may dominate at each of these scales. The variability and applicability of parameters and process formulations must be understood across the wide range of scales over which the impacts of climate change will be assessed. These techniques need to take greater advantage of remote sensing techniques and digital databases.

- Field studies in a large number of physiographic and environmental regions conducted at a range of spatial and temporal scales are needed to facilitate the two previous research needs. Studies such as FIFE, BOREAS, GCIP, etc., have provided the basis for many of the recent advances in atmospheric and hydrological models. However, these represent only a limited number of regions. In addition, they have been much too short in duration to permit studies of their responses to the broad range of climate variability associated with these regions.

- Quantitative measures of uncertainty in model parameters and model results are needed. Uncertainty measures could provide estimates of the confidence limits on model results and would be of value in the applications of these results in risk and policy analyses. The possible effects of a cascade of errors through coupled models also need to be evaluated in order to provide comparable measures of uncertainty for coupled systems.

- Improved methodologies to develop climate change scenarios are needed. Removing the uncertainties in current scenarios is dependent on improvements in both GCMs and scenario generation procedures. Scenarios must provide the spatial and temporal resolution required by assessment models, and they must incorporate the simulated changes in the means and variability of the climate variables. A more consistent set of scenario development procedures is also needed to facilitate comparisons of assessments for different regions.

- Simulation capabilities have generally exceeded the available databases. Detailed data sets for a variety of climatic and physiographic regions, collected at a range of spatial and temporal scales, are essential both for improving our understanding of hydrological processes and for testing and validating the more physically based models that are now being developed.

9 Conclusions and recommendations

M. BONELL, J. C. VAN DAM and V. JONES

This final report of the Working Group of UNESCO's IHP-IV project H-2.1 provides an overview of ongoing activities in the 'trans-science' domain that is the concern of hydrologists, water resource engineers, climatologists and meteorologists. The members of the first two communities need inputs from the latter two, but climate change and variability is in turn determined by changes in hydrological conditions. The interaction between climate and hydrological conditions is also affected, either directly or indirectly, by other factors such as population pressure and changes in land use.

This chapter presents the final conclusions of the Working Group, and their recommendations for research and measures in the future. It is hoped that the implementation of these recommendations will help to improve the practical utility of the results of large-scale modelling at the catchment scale for water resource managers and policy makers.

9.1 CONCLUSIONS

The conclusions of the Working Group are discussed under the following headings:

(a) the use of paleoclimate scenarios;
(b) the use of GCM scenarios;
(c) hydrological models and climate change;
(d) the uncertainties associated with GCMs and hydrological models;
(e) field experiments to improve GCMs;
(f) climate change and water resources management; and
(g) the outcomes of the IPCC process.

(a) The use of paleoclimate scenarios

Paleoclimate scenarios are used to provide analogues of changes in climate that have actually occurred in the past, even though the causal factors may not be known and therefore can not be easily incorporated into GCMs. The out-comes of paleoclimate change scenarios could therefore complement the scenarios from GCMs. For example, three epochs – the Holocene, the Pleistocene and the Pliocene – could be used to represent the effects of global temperature changes. According to Shiklomanov (Chapter 2), greater attention should be paid to the manipulation of such paleoclimate scenarios within the IPCC process.

Sircoulon *et al.* (Chapter 6), for example, describe the shifts in climate that have occurred in the African Sahel in historical times. However, the results of paleoclimate reconstructions often demonstrate inconsistencies with the outputs of GCMs for the region.

(b) The use of GCM scenarios

From the survey of the results of GCMs for the various world regions presented in Chapters 3–7, several main features of the impacts of climate change and variability on hydrological regimes can be identified.

In *cold and temperate regions,* the most significant changes in hydrological regimes are likely to occur in the seasonal distribution of runoff, rather than in mean annual values. There is general consensus (e.g. Shiklomanov, Arnell, Leavesley) that the greatest increases in runoff would occur in winter, due to the combined effects of higher temperatures reducing snow cover, and increases in the frequency of occurrence and the intensity of storms. Such effects would be most acute in regions where at present river regimes are dominated by spring snowmelt, especially at high northern latitudes and in mountainous areas. Thus there would be a shift in the frequency of floods and low flows, with such extrema occurring more often in winter and summer, respectively.

Shiklomanov and Shiklomanov (Chapter 7), however, indicate possible exceptions to these trends, noting that an increase in air temperature would have a greater effect on runoff than changes in precipitation, particularly in cold regions. Long-term hydrometeorological data for the

Yenisey River system in Russia have shown that maximum warming has occurred in the southern part of the basin, and minimum warming in the north, on the Arctic Ocean coast. These observations are inconsistent with the results of both GCM and paleoclimate scenarios, all of which have forecast greater warming at higher latitudes.

Many efforts have been made to evaluate the effects of climate change on hydrological regimes in temperate regions using GCM scenarios. However, according to Arnell (Chapter 5), the results of the studies undertaken within Europe are very difficult to compare because they have used different methodologies, and different climate change scenarios. Better coordinated investigations using similar scenarios across the whole of Europe are required before any firm conclusions can be drawn.

There is general agreement that the *arid and semi-arid regions* are particularly vulnerable to climate change (Shiklomanov, Sircoulon *et al.*, Leavesley), such that any changes in precipitation patterns will have significant impacts on river regimes. Using the examples of the Sahel and the Nile basin, Sircoulon *et al.* describe how runoff in Africa is much more sensitive to changes in rainfall than to changes in potential evaporation; reduced precipitation would lead to reduced vegetation cover (aggravated by human activities), which in turn would lead to reduced evaporation. Sircoulon *et al.* believe that this mechanism has been insufficiently incorporated within the present generation of GCMs.

In the *humid tropics and monsoon regions* of Asia, Shiklomanov and Shiklomanov (Chapter 7) note that the outputs of different GCMs are often contradictory, especially for monsoon climates. Nonetheless, with global warming of 1–2°C, the total runoff in both Asia and Australia (including the wetter tropical areas) is likely to increase.

Several authors note a number of shortcomings in the capabilities of the present generation of GCMs. For example, they do not take into account a number of large-scale phenomena, such as the negative (cooling) effects of aerosols in the atmosphere (with the exception of the recent work of Mitchell *et al.*, 1995), and are unable to forecast future changes in, and the likely impacts of the El Niño–Southern Oscillation (ENSO) and the preferred tracks of tropical cyclones. It is now well established that these last two phenomena have major impacts on the present-day hydrology of most of the tropics (e.g. the Amazon basin; see Chapter 3) and some temperate regions. Until such phenomena can be modelled with confidence, then in this respect the outputs of GCMs will be of little value to hydrologists.

There is a great deal of uncertainty in the GCM results on the impacts of climate change on groundwater and water quality, indicating that further research is required. According to Arnell, climate change could bring about changes in the rates of natural groundwater recharge, either directly, via changes in precipitation and total evaporation (evaporation/transpiration), or indirectly, due to changes in land surface conditions (infiltration capacity). So far, such effects have barely been studied, yet changes in groundwater levels will affect both the base flows of streams and water quality (due to saltwater intrusion, for example). The GCM results for water quality are particularly uncertain because this will be linked with future socio-economic developments and demographic changes. Complicating factors include the changing rates of groundwater abstraction, both in total and in the seasonal distribution in response to changing demands due to climate change. Once again, the arid and semi-arid regions would be the most vulnerable. In view of the lack of attention being given to groundwater recharge, and the continued uncertainty associated with changes in land use, more attention needs to be given to land surface parameterization in GCMs, especially surface soil hydraulic properties, to determine the relation between infiltration and runoff. The capability to monitor the impacts of climate variability and change are also hindered by major deficiencies in the global water quality database (Global Environmental Monitoring System, GEMS-Water, Canada).

Arnell notes that there have been no assessments of the relative importance of changes in land use induced by socio-economic activities, *vis à vis* those due to climate change. At the local scale, in some temperate regions the hydrological impacts of changes in land use may be more important than the corresponding impacts of climate change.

(c) Hydrological models and climate change

In his comprehensive review of the status of hydrological modelling with respect to climate change, Leavesley has identified a number of common problem areas in the application of a variety of models. These problems are related to modelling issues such as parameter estimation, the temporal and spatial scales of application, validation, climate scenario generation, the availability of data, and the modelling tools. The capability of models to assess the effects of climate change would be considerably improved if these problems could be resolved. To address these problems, further research is required:

(i)   Hydrological models should incorporate a more physically based understanding of hydrological processes and their interactions. The complexities of hydrological systems are such that process parameterizations will always represent the integration of the spatial heterogeneity of the factors that control these processes. However, when these parameterizations are based on the physics of the processes, the ability to measure or estimate parameter values from climate and basin characteristics is improved. Only through the use of parameterizations that do not require calibration will the problems of climatic and geographic transferability be resolved. At this stage it is not possible to judge which hydrological models are the most appropriate for coupling with GCMs. As indicated by Leavesley, there are still some unresolved issues associated with the more complex group of models which are attempting to be more physically based. One such issue is the inappropriate use of algorithms (e.g. infiltration equations) that have been developed at scales much smaller than the scale of interest.

(ii) Parameter measurement and estimation techniques must be developed for applications over a range of spatial and temporal scales. In moving from points to hillslopes to grid cells or small basins, different sets of physical laws may dominate at each of these scales. The variability and applicability of parameters and process formulations must be understood across the wide range of scales over which climate change impacts will be assessed.

(d) The uncertainties associated with GCMs and hydrological models

Even under a stationary climate, the future impacts of non-climate factors, such as socio-economic activities, technological developments and population growth, on land use and hydrology already represent major areas of uncertainty. In GCM-based climate change scenarios, such uncertainties are further accentuated by two sources of 'noise': (i) the noise in the GCM results, which gives rise to different results even for the same data set; and (ii) the noise in the form of the considerable errors that arise in the use of different hydrological methods to translate the outputs of GCM-based scenarios to water resource characteristics. After more than 15 years of research on this subject, the progress in reducing both of these sources of noise has been disappointing.

During the review of the IHP's work with the external assessors, it was generally agreed that the problem of the noise in GCM results can only be resolved by the meteorologists; the hydrological community cannot take direct

action, except to bring to the attention of GCM modellers the need for improved land surface parameterization of surface hydrology. Improved methodologies are needed, coupled with substantially greater standardization of the techniques that are used to interpret the results to water resources. Improvements are needed in both GCMs and the scenario generation procedures. Scenarios must provide the spatial and temporal resolution required by hydrological models, and should incorporate simulated changes in the means and variability of climate variables. The hydrological community could be more pro-active by continually improving hydrological modelling through field validation, linked with comparisons of a number of different models using common data sets for different climatic regions.

(e) Field experiments to improve GCMs

A number of multidisciplinary field experiments have been designed to help resolve the uncertainties connected with GCM outputs. The results of these experiments, both ongoing and planned (see Chapters 3–6), will be used to improve the land surface parameterization of GCMs, including from the hydrological perspective. However, with the possible exception of FIFE, most of these field experiments have suffered from two major disadvantages. First, they have tended to focus on the vertical terrestrial–atmospheric exchanges of energy and water vapour, and until recently there have been few detailed studies of the horizontal transfers of water fluxes. Second, they have been short term in duration, so that longer-term monitoring has been neglected. Such deficiencies have been the direct result of limitations of financial support and of the availability of scientists.

Large-scale experiments such as GEWEX, HAPEX, etc. are now under way to increase the knowledge base necessary for coupling atmospheric dynamics with hydrological process models. The design of such experiments is far from ideal, but is improving with experience. An important step has been the adoption of a catchment approach rather than to use grid squares with rectangular boundaries, but further studies of the lateral transport of energy and water vapour across boundaries are required.

In the implementation of the GEWEX Continental-scale International Project (GCIP) and the Large-scale Biosphere–Atmosphere Experiment in Amazonia (LBA), a catchment approach will be adopted. The LBA will incorporate a nested drainage basin approach (NDBA) in order to address the problem of scaling from the micro-, through the meso-, to the macroscale. The NDBA concept is the product of a formal agreement between UNESCO's IHP and IGBP–BAHC.

Both the GCIP and the LBA will operate beyond the minimum requirement of five years (Sircoulon *et al.*, Chapter 6), and will incorporate intensive observation periods, supplemented by medium- and even longer-term monitoring (with less instrumentation) appropriate to the time behaviour of each process, and to the climate conditions of each zone.

The FRIEND programme could provide an alternative strategy to address the problem of the uncertainties that seem to be inherent in climate change scenarios. Within FRIEND, sets of high-quality data on runoff regimes in Europe are being collected, and will allow more detailed analyses to detect hydrological variability, including possible linkages with atmospheric and sea surface temperature anomalies.

(f) Climate change and water resources management

As noted by Shiklomanov (Chapter 2), because of the considerable uncertainty arising from the inconsistencies in the outputs from GCMs, water resource managers face equally high uncertainty in their task of formulating appropriate adaptive strategies. Such choices of strategy are further complicated by current and future demographic trends, allied with socio-economic activities. In areas of rapid population growth, particularly in developing countries, the available water per capita will decrease, irrespective of climate change.

There is general consensus (Shiklomanov, Sircoulon *et al.*) that the arid and semi-arid regions are the most vulnerable to changes in temperature and precipitation. In such areas, any changes in runoff regimes in the larger shared river basins due to climate change could lead to international disputes. At present, the spatial and temporal resolution of GCMs do not offer the drainage basin-specific information required to provide robust estimates of water availability linked with climate change. There have been few such studies, except for the USA.

(g) The outcomes of the IPCC process

In 1988 the Intergovernmental Panel on Climate Change (IPCC) was established to gather together and assess the results of research that has been conducted around the world on various aspects of climate variability and change. The Panel was then to use that information to evaluate the environmental and socio-economic consequences of climate change, to formulate realistic response strategies, and, above all, to publicize their findings to the scientific community, international organizations and decision-makers.

The IPCC has fulfilled most of these objectives, and in this respect the process has been beneficial and useful to the international community in general. Nonetheless, the IPCC has not been able to recommend one future climate change scenario or resulting methodological approach for all assessments of the likely consequences of climate change. In the context of water resources, the outcomes of the IPCC exercise will not be practical, unless and until the predictive process can capture realistically the global processes that are known to impact on water resources, such as the El Niño–Southern Oscillation, ENSO; the tracks of tropical cyclones; and sea surface temperature anomalies linked with the Sahelian drought. These processes influence climate variability at the seasonal, annual and decadal time scales that are of interest to hydrologists.

9.2 RECOMMENDATIONS

From the survey of the work of the IPCC, and from the analysis of the studies conducted in the various continents as described in the present report, it appears that little progress has been made over the last six years in improving the reliability of forecasts of climate change, and of the probable impacts of such changes on water resources, the environment and society. The Working Group therefore proposes the following recommendations.

1. *Coordination*: The many studies of the likely impacts of climate change and variability on hydrological regimes that have been conducted thus far have used a wide variety of scenarios and models, so that the results are not always directly comparable. In future, climate change scenarios and models should be selected so that the results will be directly comparable. This selection should be based on a coordinated scientific analysis by the international community of modellers of climate and hydrology.

2. *Forecasts of future conditions*: Policy makers require appropriate information, at the catchment scale, that will enable them to make sound decisions on the most effective action plans to mitigate or adapt to the new conditions resulting from climate change. There is therefore a need for more accurate forecasts of increases in greenhouse gases in the atmosphere, and of changes in land use. Such forecasts require international cooperation.

3. *Adaptive measures*: As trends in climate become clearer, they may justify plans for adaptive measures relating

not only to hydrological conditions and water management, but also to other areas such as land use and agricultural practices. Policy makers, planners and water resource engineers should be alert to such trends in order that timely action can be taken.

4. *Sensitivity analysis:* The higher the hydrological sensitivity of a region, the more attention should be given to more accurate determination of changes in climate variables, subject to the availability of and demand for water resources. Policy makers and planners should therefore bring their concerns to the attention of climatologists.

5. *Water resource management studies*: UNESCO-IHP should frame a programme of water resource management studies and applications, including as inputs the results of GCMs and subsequent analyses of data on climate change and variability.

6. *Anticipating extreme events*: Several studies have indicated that in some regions extreme events such as floods and droughts will become more severe and/or more frequent, and so merit greater attention. For the design, adaptation and operation of water infrastructure projects in these regions, water resources engineers will need to anticipate future regimes. For vulnerable locations, it may be useful to consider the transposition of existing rainfall intensity–frequency–duration curves from other locations that are representative of the predicted changes in climate. This would be particularly relevant for urban areas, because of their vulnerability to flooding in the event of intense rainfall over relatively small areas. Water resources engineers should also pay greater attention to regions where precipitation is likely to decrease.

7. *Large-scale field experiments and modelling approaches*: Large-scale field experiments have so far focused on the vertical terrestrial–atmospheric exchanges of energy and water vapour, rather than on details of horizontal water fluxes. In future, the organizers of such experiments should pay greater attention to quantifying horizontal water fluxes in order to improve the effectiveness of the nested drainage basin approach (NDBA).

8. *Future large-scale field experiments*: Decisions as to whether new large-scale field experiments should be set up in other climatic regions should be taken after weighing the value of the expected results (not only for hydrologists and water resource managers) against the inevitably high costs of such experiments. Except for some large countries, these decisions should be made by international agreements and with joint funding.

9. *Freely accessible databases*: Future large-scale studies should include the compilation of complete and freely accessible databases, as is currently being done within HAPEX-Sahel.

10. *Data collection network:* In order to be able to study the effects of changes in both climate and in land use, it is important to have access to long-term data sets compiled both in the past and in the future. The World Meteorological Organization and individual nations should therefore recognize that a global network of climate and hydrological monitoring stations needs to be maintained at an adequate density. This is particularly the case in areas where data collection has been discontinued due to natural disasters such as floods or droughts, or to unstable political situations, which may have resulted in the disruption of public services.

11. *Groundwater*: There are still considerable uncertainties concerning the impacts of climate change on groundwater recharge. It is recommended that an inventory is made of the studies that have been undertaken so far of the impacts of climate change on groundwater regimes. Only then should areas for further study be selected, and appropriate research proposals formulated. The selected areas should be located in different climatic zones, and the proposed groundwater studies linked to ongoing studies of other components of the hydrological cycle in those areas. Such an inventory and subsequent research proposals might be undertaken as a UNESCO-IHP project.

12. *Funding*: At the international level, many countries have acknowledged their concerns about the possible impacts of climate change by signing the Framework Convention on Climate Change (FCCC) and by supporting the IPCC process. At the national level, however, financial support for long-term field experiments has been declining. To resolve this apparent paradox, and to ensure more effective water policy decision-making, it is essential to maintain field experiments at the catchment scale to monitor and evaluate the impacts of climate change. International funding agencies should focus on areas for which the data are scarce, and where the most serious problems, due to hydrological conditions and/or human impacts, are expected. North Africa, particularly the Sahelian region, should be given high priority. Because the process of climate change is slow, and water resources engineers need to pay particular attention to extremes,

ongoing and new field experiments need to be established for periods of decades at least. Successful modelling of the impacts of climate change requires long-term experiments in order both to ensure that the broadest ranges of conditions are considered, and to maximize the returns on financial investments.

References

Abbott, M.B, Bathurst, J.C., Cunge, J.A., O'Connell, P.E. and Rasmussen, J. (1986) An introduction to the European Hydrological System, Système Hydrologique Européen, SHE, 1: History and philosophy of a physically-based, distributed modeling system. *J. Hydrology* **87**, 45–59.

Albergel, J. (1987) *Sécheresse, Désertification et Ressources en Eau de Surface. Application aux Petits Bassins du Burkina Faso.* IAHS Publ. No. 168, pp. 355–65.

Albergel, J., Carbonnel, J.-P. and Grouzis, M. (1985) *Pejoration climatique au Burkina Faso: Incidences sur les ressources en eau et les productions végétales.* Cah. ORSTOM, sér. Hydrology **XXI**(1), 3–19.

Albergel, J. and Gioda, A. (1986) *Extensions des surfaces agricoles et modification de l'écoulement. Analyse sur deux bassins de la savane africaine.* 10èmes Journées de l'Hydraulique, SHF Rep. No. 9, Paris.

Alberty, R., Crum, T. and Toepfer, F. (1991) The NEXRAD Program: Past, present, and future, A 1991 perspective. In *Proceedings of an International Conference on Radar Meteorology.* American Meteorology Society, June 24–28.

Alley, W.M. (1984) On the treatment of evapotranspiration, soil moisture accounting and aquifer recharge in monthly water balance models. *Water Resources Res.* **20**, 1137–49.

Anderson, E.A. (1973) *National Weather Service River Forecast System: Snow Accumulation and Ablation Model.* NOAA Technical Memorandum NWS Hydro-17, US Dept. of Commerce, Silver Springs, MD.

André, J.C., Goutorbe, J.P. and Perrier, A. (1986) A hydrologic atmospheric experiment for the study of water budget and evaporation flux at the climate scale. *Bull. Am. Meteorol. Soc.* **67**, 138–44.

André, J.C. *et al.* (1988) Evaporation over land surfaces: first results from HAPEX-MOBILHY special observing period. *Ann. Geophys.* **6**, 477–92.

Anthes, R.A., Hsie, E.Y. and Kuo, Y.H. (1987) *Description of the Penn State/NCAR Mesoscale Model Version 4 (MM4).* Tech. Note NCAR/ TN-282+STR, National Center for Atmospheric Research, Boulder, CO.

Arnell, N.W. (1992a) Factors controlling the effects of climate change on river flow regimes in a humid temperate environment. *J. Hydrology* **132**, 321–42.

Arnell, N.W. (1992b) Impacts of climate change on river flow regimes in the UK. *J. Inst. Water and Environmental Management* **6**, 432–42.

Arnell, N.W. (1994) Variations over time in European hydrological behaviour: A spatial perspective. In *FRIEND: Flow Regimes from International Experimental and Network Data.* IAHS Publ. No. 221, pp. 179–84.

Arnell, N.W. (1995) *Impact of Climate Change on Hydrological Regimes and Water Resources in the European Community.* First Progress Report EV5V-CT93-0293. Institute of Hydrology, Wallingford, UK.

Arnell, N.W., Brown, R.P.C. and Reynard, N.S. (1990) *Impact of Climatic Variability and Change on River Flow Regimes in the UK.* Report 107, Institute of Hydrology, Wallingford, UK.

Arnell, N.W., Gottschalk, L., Krasovskaia, I. and van der Wateren, B. (1993) Large-scale variations in hydrological characteristics across Europe. In A. Gustard (Ed.) *Flow Regimes from International Experimental and Network Data (FRIEND).* Institute of Hydrology, Wallingford, UK, pp. 112–37.

Arnell, N.W. and Reynard, N.S. (1993) *Impact of Climate Change on River Flow Regimes in the United Kingdom.* Institute of Hydrology Report to the Department of the Environment.

Arnell, N.W. and Reynard, N.S. (1995) *Impact of Climate Change on Water Resources in East Africa.* Institute of Hydrology Report to Overseas Development Administration, UK.

Assel, R.A. (1991) Implications of CO_2 global warming on Great Lakes ice cover. *Climatic Change* **18**, 377–95.

Aston, A.R. (1984). The effect of doubling atmospheric carbon dioxide on streamflow: A simulation. *J. Hydrology* **67**, 273–80.

Australian Bureau of Meteorology (1991) *The Impacts of Climate Change on Hydrology and Water Resources.* Climate Change Impacts Workshop, Melbourne, June 1991.

Avissar, R. and Pielke, R.A. (1989) A parametrization of a heterogeneous land surface for atmospheric numerical models and its impact on regional meteorology. *Monthly Weather Rev.* **117**, 2113–36.

Avissar, R. and Verstraete, M.M. (1990) The representation of continental surface processes in atmospheric models. *Rev. Geophys.* **28**, 35–52.

BALTEX (1995) *Baltic Sea Experiment, BALTEX: Initial Implementation Plan.* International BALTEX Secretariat, Geesthacht, Germany.

Bao, W.M. (1994) A conceptual flow-sediment coupled simulation model for large basins. *Advances in Water Science* **5**(4) (in Chinese).

Barros, V., Castañeda, M.E. and Doyle, M. (1995) *Recent Precipitation Trends in Southern South America to the East of the Andes: An Indication of a Mode of Climate Variability.* Proc. Workshop e Seminario Latino Americano sobre Emissões Mundiais de Gases de Efeito Estufa no Setor Energético e seus Impactos. July, COPPE, UFRJ, Rio de Janeiro.

Barry, R.G. and Chorley, R.J. (1987) *Atmosphere, Weather and Climate.* Routledge, London.

Barth, H. (1860) *Voyages et Découvertes dans l'Afrique Septentrionale et Centrale Pendant les Années 1849 à 1855,* 4 vols. Bohné, Paris.

Bates, B.C. *et al.* (1995) *Impact of Climate Change on Australia's Surface Water Resources.* Proceedings Greenhouse '94 Conferemce, October 1994, Wellington, New Zealand.

Bathurst, J.C. and O'Connell, P.E. (1992) Future of parameter modelling: The Système Hydrologique Européen. *Hydrological Processes* **6**, 265–77.

Becker, A. and Serban, P. (1990) *Hydrological Models for Water Resources System Design and Operation.* Operational Hydrology Report No. 34, WMO No. 740, Geneva.

Bell, B. (1971) The Dark Ages in ancient history: The first Dark Age in Egypt. *Am. J. Archaeology* **75**, 1–36.

Bergstrom, S. (1976) *Development and Application of a Conceptual Runoff Model for Scandinavian Catchments.* Bulletin Series A-52, Dept. of Water Resources Engineering, Lund Institute of Technology, Swedish Meteorological and Hydrological Institute (Norrköping, Sweden).

Berndtsson, R., Larson, M., Lindh, G. *et al.* (1989) Climate-induced effects on the water balance: Preliminary results from studies in the

Varpinge experimental research basin. In *Proceedings of a Conferemce on Climate and Water*. Helsinki, Finland, Vol. 1, WMO, Geneva, pp. 437–49.

Beven, K. (1985) Distributed models. In M.G. Anderson and T.P. Burt (Eds.) *Hydrological Forecasting*. Wiley, Chichester.

Beven, K. (1989) Changing ideas in hydrology: The case of physically based models. *J. Hydrology* 105, 157–72.

Beven, K. and Binley, A. (1992) The future of distributed models: Model calibration and uncertainty prediction. *Hydrological Processes* 6, 279–98.

Blyth, E.M., Dolman, A.J. and Noilhan, J. (1994) The effect of forest on mesoscale rainfall: An example from HAPEX-MOBILHY. *J. Appl. Meteorol.* 33, 445–54.

Bolle, H.-J., André, J.C., Arrue, J.L. *et al.* (1993) EFEDA: European Field Experiment in a Desertification-Threatened Area. *Ann. Geophys.* 11, 173–89.

Bonell, M. (1998) Possible impacts of climate variability and change on tropical forest hydrology. WWF Conference on the potential impacts of climate change on tropical forest ecosystems, Puerto Rico, April 1995. *Climatic Change*, 39, 215–72.

Bonell, M. and Balek, J. (1993) Recent scientific developments and research needs in hydrological processes of the humid tropics. In M. Bonell, M.M. Hufschmidt and J. S. Gladwell (Eds.) *Hydrology and Water Management in the Humid Tropics: Hydrological Research Issues and Strategies for Water Management*. UNESCO/Cambridge University Press, Cambridge, pp. 167–260.

Brandsma, T. (1995) Hydrological Impacts of Climate Change. PhD thesis, Delft University of Technology, Delft, The Netherlands.

Brink, H.M. ten (1996) Reduction of solar radiation by aerosols in Europe: 'Dust particles' compensate for the greenhouse effect. *Climatic Change* 30, 13–14.

Budyko, M.I. (1948) *Evaporation under Natural Conditions*. Hydrometeoizdat, Leningrad.

Budyko, M.I. (1972) *The Influence of Man on Climate*. Hydrometeoizdat, Leningrad.

Budyko, M.I. (1988) Climate at the end of the twentieth century, *Meteorologiya i Hydrologiya*, No. 10 (in Russian).

Budyko, M., Borzenkova, I., Menzhulin, G. and Shiklomanov, I., eds. (1994) *Gambios Antropogenicos del Climate en America del Sur*. Academia Nacional de Agronomia y Veterinaria.

Bultot, F., Coppens, A., Dupriez, G.L., Gellens, D. and Meulenberghs, F. (1988) Repercussions of a CO_2-doubling on the water cycle and on the water balance: A case study for Belgium. *J. Hydrology* 99, 319–47.

Bultot, F. and Dupriez, G.L. (1976) Conceptual hydrological model for an average-sized catchment area, I: Concepts and relationships. *J. Hydrology* 29, 251–72.

Bultot, F., Gellens, D., Schädler, B. and Spreafico, M. (1994) Effects of climate change on snow accumulation and melting in the Broye catchment (Switzerland). *Climatic Change* 28, 339–63.

Bultot, F., Gellens, D., Spreafico, M. and Schädler, B. (1992) Repercussions of CO_2 doubling on the water balance: A case study in Switzerland. *J. Hydrology* 137, 199–208.

Burgos, J.J., Ponce, H.F. and Molion, L.C.B. (1991) Climate change prediction for South America. *Climatic Change* 18, 223–39.

Burnash, R.J.C., Ferral, R.L. and McGuire, R.A. (1973) *A Generalized Streamflow Simulation System: Conceptual Modeling for Digital Computers*. US Dept. of Commerce, National Weather Service, and State of California, Dept. of Water Resources, Sacramento, CA.

Canadian GEWEX Science Committee (1992) *Canadian GEWEX Programme: A Conceptual Overview*, Ed. G. McBean. National Hydrology Research Centre, Saskatoon, Saskatchewan.

Campos, S. (1989) *Management of the Zambezi Basin: Social, Political and Economic Considerations*. Working Paper WP89-092, IIASA, Laxenburg, Austria.

Carter, T.R., Parry, M.L., Harasawa, H. and Nishioka, S. (1994) *IPCC Technical Guidelines for Assessing Climate Change Impacts and Adaptations*. Dept. of Geography, University College, London.

Casenave, A. and Valentin, C. (1990) *Les États de Surface de la Zone Sahélienne: Influence sur L'infiltration*. Coll. Didactiques, ORSTOM, Paris.

CCIRG (Climate Change Impacts Review Group) (1991) *The Potential Impacts of Climate Change on the United Kingdom*. HMSO, London.

CGER (1993) *The Potential Effects of Climate Change in Japan*. Center for Global Environmental Research, Tsukuba, Japan.

Chang, L.H., Hunsaker, C.T. and Draves, J.D. (1992) Recent research on effects of climate change on water resources. *Water Resources Bull.* (AWRA) 28(2), 273–86.

Changnon, S.A. and Huff, F.A. (1991) Potential effects of changed climates on heavy rainfall frequencies in the midwest. *Water Resources Bull.* 27(5), 753–9.

Charney, J.G. (1975) Dynamics of deserts and drought in the Sahel. *Q. J. R. Meteorol. Soc.* 101, 19–202.

Chunzhen, L. (1989) *Study of Climate Change and Water Resources in Northern China*. Chinese Ministry of Water Resources (unpublished).

Close, A.F. (1988) Potential impact of the greenhouse effect on the water resources of the River Murray. In G. Pearman (Ed.) *Greenhouse: Planning for Climate Change*. CSIRO, Australia, pp. 312–23.

Cohen, S.J. (1986) Impacts of CO_2-induced climatic change on water resources in the Great Lakes basin. *Climatic Change* 8, 135–53.

Cohen, S.J. (1991) Possible impacts of climatic warming scenarios on water resources in the Saskatchewan River sub-basin, Canada. *Climatic Change* 19, 291–317.

Cohen, S.J. and Allsopp, T.R. (1988) The potential impacts of a scenario of CO_2-induced climatic change on Ontario, Canada. *J. Climate* 1, 669–81.

Cohen, S.J., Walsh, L.E. and Louie, P.Y.T. (1989) *Possible Impacts of Climate Warming Scenarios on Water Resources in the Saskatchewan River Subbasin*. Canadian Climate Centre Rep. No. 89-9, Atmospheric Environment Service, National Hydrology Research Centre, Saskatoon, Canada.

COHMAP (Cooperative Holocene Mapping Project) (1988) Climatic changes of the last 18 000 years: Observations and model simulations. *Science* 241, 1043–52.

Collins, D.M. (1989) Influence of glacierisation in the response of runoff from Alpine basins to climate variability. In *Conference on Climate and Water*, Helsinki, Finland, Vol. 1, WMO, Geneva, pp. 319–28.

Conway, D. (1993) The Development of a Grid-Based Hydrologic Model of the Blue Nile and the Sensitivity of Nile River Discharges to Climate Change. PhD Thesis, University of East Anglia, UK (unpublished).

Conway, D. (1994) *The Implications of Future Climate Change for Water Resources in the Countries of the Nile Basin*. African Studies Association of the UK, Biennial Conference, University of Lancaster, September 1994.

Conway, D. and Hulme, M. (1993) Recent fluctuations in precipitation and runoff over the Nile sub-basins and their impact on Main Nile discharge. *Climatic Change* 25, 127–51.

Cooley, K.R. (1990) Effects of CO_2-induced climatic changes on snowpack and streamflow. *Hydrological Sci.* 35(5), 511–22.

Cooper, D.M., Wilkinson, W.B. and Arnell, N.W. (1995) The effect of climate change on aquifer storage and river baseflow. *Hydrol. Sci. J.* 40, 615–31.

Courtier, P. and Geleyn, J.F. (1988) A global numerical weather prediction model with variable resolution: Application to the shallow water equation. *Q. J. R. Meteorol. Soc.* 114, 1321–46.

Crane, R.G. (1992) General circulation model studies of global warming. In S.K. Majumdar, L.S. Kalkstein, B.M. Yarnal, E.W. Miller and L.M. Rosenfeld (Eds.) *Global Climate Change: Implications, Challenges, and Mitigation Measures*. Pennsylvania Academy of Science, Easton, PA, pp. 189–208.

D'Angelo, A. and Damazio, J.M. (1993) Modelo SIMBAH: modelagem da transpiração e da intercepção vegetal. *Proceedings of the 10th Brazilian Water Resources Symposium*, ABRH, Gramado, pp. 351–60 (in Portuguese).

Desbois, M. (1994) TROPIQUES, a small satellite for the study of the variability of water and energy cycles in the intertropical band. *Proceedings of a European Symposium on Satellite Remote Sensing*, Rome, September 1994, SPIE–EUROPTO series.

Dettinger, M.D. and Cayan, D.R. (1995) Large-scale atmospheric forcing of recent trends toward early snowmelt runoff in California. *J. Climate* 8, 606–23.

Dickinson, R.E. (1984) Modelling evapotranspiration for three-dimensional global climate Models. In *Climate Processes and Climate Sensitivity*. Geophysics Monographs, LF29, AGU, pp. 58–72.

Dickinson, R.E. (1988) Climate and hydrologic systems. In *Toward an Understanding of Global Change*. National Academy Press, Washington, DC, pp. 107–32.

Dickinson, R.E. and Kennedy, J. (1992) Impacts on Regional Climate of Amazon Deforestation. *Geophys. Res. Lett.* 19, 1947–50.

Dickinson, R.E., Lean, J. and Warrilow, D.A. (1989) Climate impact of Amazon deforestation. *Nature* 342, 311–13.

Dickinson, R.E. and Rowntree, P. (1993) GCM model studies. In P. Sellers, C. Nobre, D. Fitzjarrald, P. Try and D. Lucid (Eds.) *A Preliminary Science Plan for a Large-Scale Biosphere Atmosphere Field Experiment in the Amazon Basin*. ISLSCP/GEWEX, Washington, DC.

Dolman, A.J., Kabat, P., Elders, J.A., Bastiaansen, W.G.M. and Ogink-Hendriks, M.J. (1995) Regionalization and parametrization of hydrological processes at the land surface. In J.F.Th. Schulte *et al*. (Eds.) *Scenario Studies for the Rural Environment*. Kluwer, Dordrecht.

Dooge, J.C.I. (1992) Hydrologic models and climate change. *J. Geophys. Res.* 97(D3), 2677–86.

Dozier, J. (1992) Opportunities to improve hydrologic data. *Rev. Geophys.* 30(4), 315–31.

Dracup, J.A. and Kahya, E. (1994) The relationships between U.S. streamflow and La Niña events. *Water Resources Res.* 30(7), 2133–41.

Dubief, J. (1953) *Essai sur l'hydrologie superficielle au Sahara*. Dir. Serv. Col. Hydraul. Alger.

Dunne, T. (1983) Relation of field studies and modeling in the prediction of storm runoff. *J. Hydrology* 65, 25–48.

Dunne, T. and Barker, M. (1997) *The NDBA (Nested Drainage Basin Approach) in the Amazon Basin* (in preparation).

Easterling, W.E., McKenney, M., Rosenberg, N.S. and Leman, K. (1991) *A Farm Level Simulation of the Effects of Climate Change on Crop Productivity in the MINK Region*. Working Paper 11–13, US Department of Energy.

Easterling, W.E., Rosenberg, N.J., McKenney, M.S. and Jones, C.A. (1992a) An introduction to the methodology, the region of study, and a historical analog of climate change. *Agric. Forest Meteorol.* 59, 3–15.

Easterling, W.E., Rosenberg, N.J., McKenney, M.S., Jones, C.A., Dyke, P.T. and Williams, J.R. (1992b) Preparing the erosion productivity impact calculator (EPIC) model to simulate crop response to climate change and the direct effects of CO_2. *Agric. Forest Meteorol.* 59, 17–34.

Engman, E.T. and Gurney, R.J. (1991) *Remote Sensing Hydrology*. Chapman & Hall, London.

Espenshade, E.B., Jr and Morrison, J.L. 1978 *Goode's World Atlas*. Rand-McNally, Chicago.

Evans, T. (1994) History of Nile flows. In P.P. Howell and J.A. Allan (Eds.) *The Nile: Sharing a Scarce Resource*. Cambridge University Press, Cambridge, pp. 27–63.

Famiglietti, J.S. and Wood, E.F. (1994a) Multiscale modeling of spatially variable water and energy balance processes. *Water Resources Res.* 30, 3061–78.

Famiglietti, J.S. and Wood, E.F. (1994b) Application of multiscale water and energy balance models of a tallgrass prairie. *Water Resources Res.* 30, 3079–93.

Famiglietti, J.S., Wood, E.F., Sivapalan, M. and Thongs, D.J. (1992) A catchment scale water balance model for FIFE. *J. Geophys. Res.* 97(D17), 18 997–19 007.

Feddema, J.J. and Mather, J.R. (1992) Hydrological impacts of global warming over the United States. In S.K. Majumdar, L.S. Kalkstein, B.M. Yarnal, E.W. Miller and L.M. Rosenfeld (Eds.) *Global Climate Change: Implications, Challenges and Mitigation Measures*. Pennsylvania Academy of Science, Easton, PA, pp. 50–62,

Feddes, R.A. (Ed.) (1995) *Space and Time Scale Variability and Interdependencies in Hydrological Processes*. International Hydrology Series. Cambridge University Press, Cambridge.

Feddes, R.A., Menenti, M. and Kabat, P. (1989) *Modelling the Soil Water and Surface Energy Balance in Relation to Climate Models*. European Coordination Group on Land Surface Processes, Hydrology and Desertification in Europe, Barcelona.

Finkelstein, P.L. and Truppi, L.E. (1991) Spatial distribution of precipitation in the United States. *J. Climate* 4, 373–85.

Flaschka, I., Stockton, C.W. and Boggess, W.R. (1987) Climatic variation and surface water resources in the Great Lakes basin region. *Water Resources Bull.* 23(1), 47–57.

Folland, C.K., Palmer, T.N. and Parker, D.E. (1986) Sahel rainfall and worldwide sea temperatures 1901–85. *Nature* 320, 602–7.

Franchini, M. and Pacciani, M. (1991) Comparative analysis of several conceptual rainfall-runoff models. *J. Hydrology* 122, 161–219.

Gaffen, D.J., Rosen, R.D., Salstein, D.A. and Boyle, J.S. (1995) Validation of humidity, moisture fluxes, and soil moisture in GCMs: Report of AMIP diagnostic subproject 11, Part 2, Humidity and moisture flux fields. In W.L. Gates (Ed.) *Proceedings 1st International AMIP Scientific Conference*, WCRP-92, WMO/TD No. 732, WMO, Geneva, pp. 91–6.

Gan, T.Y. and Burgess, S.J. (1990) An assessment of a conceptual rainfall–runoff model's ability to represent the dynamics of small hypothetical catchments. 2. Hydrologic responses for normal and extreme rainfall. *Water Resources Res.* 26(7), 1605–19.

Gates, W.L. (1992) AMIP: The atmospheric model intercomparison project. *Bull. Am. Meteorol. Soc.* 73(12), 1962–70.

Gates, W.L. (Ed.) (1995a) *Proceedings of the 1st International AMIP Scientific Conference*, WCRP-92, WMO/TD No. 732, WMO, Geneva.

Gates, W.L. (1995b) An overview of AMIP and preliminary results. In W.L. Gates (Ed.) *Proceedings of the 1st International AMIP Scientific Conference*, WCRP-92, WMO/TD-No. 732, pp. 1–8.

Gauzer, B. (1994) Effect of temperature change on the catchment over the Nagymaros section on the flow conditions of the Danube. In *Climate Change Impacts on the Water Resources of Hungary including the Upper Danube Basin*. VITUKI, Budapest/IIASA, pp. 51–82.

Gellens, D. (1991) Impact of a CO_2-induced climate change on river flow variability in three rivers in Belgium. *Earth Surface Processes and Landforms* 16, 619–25.

Georgievsky, V.Yu. (1978) Computations and forecasts of changes in water level at the Caspian Sea under the effect of natural climatic factors and man's impact. *Trudy GGI*, 255, 94–112 (in Russian).

Georgievsky, V.Yu. and Ezhov, A.V. (1990) Prospects for joint control of water and salt balances of the Caspian Sea and Sea of Azov. In *Proceedings of the 5th All-Union Hydrological Congress*, 4, 178–85.

Georgievski, V.Yu., Shalygin, A.L. and Doganovskaya, T.M. (1993) Modern and future dynamics of crop demand for irrigation in connection with global climate warming, *Meteorology and Hydrology*, No. 12, pp. 81–7.

Ghan, S.J. (1992) The GCM credibility gap. *Climatic Change* 21, 345–6.

Ghassemi, F., Jacobsen, G. and Jakeman, A.S. (1991) Major Australian aquifers: Potential climatic change impacts. *IWRA, Water International* 16(1), 38–44.

Gilyen-Hofer, A. (1994) Impact of potential climate change on long-term discharge data series. In *Climate Change Impacts on the Water Resources of Hungary including the Upper Danube Basin*. VITUKI, Budapest/IIASA, pp. 51–82.

Giorgi, F. and Mearns, L.O. (1991) Approaches to the sanitation of regional climate change: A review. *Rev. Geophys.* 29(2), 191–216.

Gleick, P.H. (1987a) The development and testing of a water balance model for climate impact assessment: Modeling the Sacramento basin. *Water Resources Res.* 23(6), 1049–61.

Gleick, P.H. (1987b) Regional hydrologic consequences of increases in atmospheric CO_2 and other trace gases. *Climatic Change* 10, 137–61.

Gleick, P.H. (1988) Climatic change and California: Past, present and future vulnerabilities. In M. Glantz (Ed.) *Societal Response to Regional Dynamic Change: Forecasting by Analogy*. Westview Press, Boulder, CO.

Gleick, P.H. (1990) Vulnerability of water systems. In P.E. Waggoner (Ed.) *Climate Change and U.S. Water Resources*. Wiley, New York, pp. 223–40.

Gleick, P.H. (1991) The vulnerability of runoff in the Nile Basin to climatic changes. *The Environmental Professional* 13, 66–73.

Goodrich, D.C. (1994) SALSA-MEX: A large-scale semi-arid land surface atmospheric mountain experiment. In *Proceedings of the International Geoscience and Remote Sensing Symposium*, Vol. 1, Pasadena, CA, pp. 190–3.

Goutorbe, J.P., Lebel, T., Tinga, A., Bessemoulin, P., Dolman, H., Engman, E.T., Gash, J.H.C., Hoepffner, M., Kabat, P., Kerr, Y.H.,

Monteny, B., Prince, S., Saïd, F., Sellers, P. and Wallace, J. (1992) *Experimental Plan for HAPEX-SAHEL*. ORSTOM, Paris.

Goutorbe, J.P., Lebel, T., Tinga, A., Bessemoulin, P., Brouwer, J., Dolman, H., Engman, E.T., Gash, J.H.C., Hoepffner, M., Kabat, P., Kerr, Y.H., Monteny, B., Prince, S., Saïd, F., Sellers, P. and Wallace, J. (1994) HAPEX-SAHEL: A large-scale study of land–atmosphere interactions in the semi-arid tropics. *Ann. Geophys.* **12**, 53–64.

Graham, N.E. (1995) Simulation of recent global temperature trends. *Science* **267**, 666–71.

Grayson, R.B., Moore, I.D. and McMahon, T.A. (1992) Physically based hydrologic modeling, 2: Is the concept realistic? *Water Resources Res.* **26**(10), 2659–66.

Griffiths, G.A. (1989) *Water Resources in New Zealand: Report on the Impacts of Climate Change*. North Canterbury Catchment Board, New Zealand.

Grotch, S.L. (1988) *Regional Intercomparison of General Circulation Model Predictions and Historical Climate Data*. U.S. Dept. of Energy, DOE/NBB-0084.

Grotch, S.L. and MacCracken, M.C. (1991) The use of general circulation models to predict regional climatic change. *J. Climatology* **4**, 286–303.

Gustard, A. (Ed.) (1993) *Flow Regimes from International Experimental and Network Data (FRIEND)*. Institute of Hydrology, Wallingford (3 vols).

Gustard, A., Roald, L.A., Demuth, S., Lumadjeng, H.S. and Gross, R. (1989) *Flow Regimes from Experimental and Network Data (FRIEND)*. Institute of Hydrology, Wallingford (2 vols).

Halldin, S., Gottschalk, L., van de Griend, A.A. *et al.* (1995) *Science Plan for NOPEX*. NOPEX Central Office, Uppsala.

Hansen, J., Fung, I., Lacis, A., Rind, D., Russell, G., Lebedeff, S., Ruedy, R. and Stone, P. (1988) Global climate changes as forecast by the Goddard Institute for Space Studies three-dimensional model. *J. Geophys. Res.* **93**, 9341–64.

Hansen, J., Lacis, A., Rind, D., Russell, G., Stone, P., Fung, I., Ruedy, R. and Lerner, J. (1984) Climate sensitivity: Analysis of feedback mechanisms. In J.E. Hansen and T. Takahashi (Eds.) *Climate Processes and Climate Sensitivity*. Geophysical Monograph Series, vol. 29, pp. 130–63.

Hare, F.K. (1983) *Climate and Desertification: A revised analysis*. WCP44, WMO.

Hay, L.E., McCabe, Jr, G.J., Wolock, D.M. and Ayers, M.A. (1992) Use of weather types to disaggregate general circulation model predictions. *J. Geophys. Res.* **97**(D3), 2781–90.

Henderson-Sellers, A., Pitman, A.J., Love, P.K., Irannejad, P. and Chen, T.H. (1995) Project for Intercomparison of Land Surface Parameterization Schemes (PILPS): Phases 2 and 3. *Bull. Am. Meteorol. Soc.* **74**(7), 1335–49.

Henderson-Sellers, A. and Robinson, P.J. (1992) *Contemporary Climatology*. Longman, Harlow, UK.

Henderson-Sellers, A., Yang, Z.-L. and Dickinson, R.E. (1993) Project for intercomparison of land surface parameterization schemes. *Bull. Am. Meteorol. Soc.* **74**(7), 1335–49.

Hinzman, L.D. and Kane, D.L. (1992) Potential response of an Arctic watershed during a period of global warming. *J. Geophys. Res.* **97**(D3), 2811–20.

Hisdal, H., Erup, J., Gudmundsson, K. *et al.* (1995) *Historical Runoff Variations in the Nordic Countries*. Nordic Hydrological Programme Report No. 37.

Hondzo, M. and Stefan, H.G. (1993) Regional water temperature characteristics of lakes subjected to climate change. *Climatic Change* **24**, 187–211.

Hostetler, S.W. (1994) Hydrologic and atmospheric models: The (continuing) problem of discordant scales. *Climatic Change* **27**, 345–50.

Hostetler, S.W. and F. Giorgi (1993) Use of output from high-resolution atmospheric models in landscape-scale hydrologic models: An assessment. *Water Resources Res.* **29**(6), 1685–95.

Hubert, P. and Carbonnel, J.P. (1987) Approche statistique de l'aridification de l'Afrique de l'Ouest. *J. Hydrology* **95**, 165–83.

Hulme, M. (1990) Global climate change and the Nile Basin. In P.P. Howell and J.A. Allan (Eds.) *The Nile: Sharing a Scarce Resource*. Cambridge University Press, Cambridge, pp. 139–62.

Hulme, M. (1992) Rainfall changes in Africa, 1931–60 to 1961–90. *Int. J. Climatology* **12**, 685–99,

Hulme, M., Conway, D. Kelly, M.P., Subak, S. and Downing, T.E. (1994) *The Impacts of Climate Change on Africa*. Stockholm Environment Institute (SEI).

Hurst, H.E. (1951) Long-term storage capacity of reservoirs. *Trans. ASCE* **116**, 770–99.

Hurst, H.E. (1965) *Long-Term Storage*. Constable, London.

Hydrometeoizdat (1987) *Anthropogenic Climate Change*. Moscow, Hydrometeoizdat.

ICASVR (1992) *Liaison Reports on ICASVR-related Activities*. International Committee on Atmosphere, Soil and Vegetation Relations, International Association of Hydrological Sciences (IAHS), Velp, the Netherlands.

Idso, S.B. and Brazel, A.S. (1984) Rising atmospheric carbon dioxide concentrations may increase streamflow. *Nature* **312**, 51–3.

IGPO (1994) *Implementation Plan for the GEWEX Continental-Scale International Project (GCIP)*, Vol. II. International GEWEX Project Office, Washington, DC.

IHP-ICASVR-IGBP (1995) *State of the Art Report on Land Surface Processes in Regional and Large Scale Hydrology*.

IPCC (1990a) *Climate Change: The IPCC Scientific Assessment*, Eds. J.T. Houghton, G.S. Jenkins and S.S. Ephraims. Cambridge University Press, Cambridge, for WMO/UNEP.

IPCC (1990b) *Policymakers' Summary of the Scientific Assessment of Climate Change*. Report prepared by IPCC Working Group I, Geneva.

IPCC (1991) *Scientific Assessment of Climate Change: Summary of the IPCC Working Group I Report*. Proc. 2nd World Climate Conference, Eds. J.T. Houghton *et al.* Cambridge University Press, Cambridge, pp. 23–44.

IPCC (1992) *Climate Change 1992: Supplementary Report to the IPCC Scientific Assessment*, Eds. J.T. Houghton, B.A. Callender and S.K. Varney. Cambridge University Press, Cambridge.

IPCC (1995) *Second IPCC Report*, in particular Chapter 10: Hydrology and freshwater ecology, and Chapter 14: Water resources management. Cambridge University Press, Cambridge.

Issar, A.S. (1995) *Impacts of Climate Variations on Water Management and Related Socio-economic Systems*. Technical Documents in Hydrology, International Hydrological Programme, UNESCO, Paris.

Jarvis, C.S. (1936) Flood-stage records of the River Nile. *Trans. ASCE* **101**, 1012–71.

Jones, P.D. and Briffa, K.R. (1992) Global surface air temperature variations during the twentieth century, Part 1: spatial, temporal and seasonal details. *The Holocene* **2**(2), 165–79.

Jones, R.G., Murphy, R.G. and Noguer, M. (1995) Simulation of climate change over Europe using a nested regional climate model. 1: Assessment of current climate, including sensitivity to location of lateral boundaries. *Q. J. R. Meteorol. Soc.* **121**, 1413–49.

Kaczmarek, Z. (1990a) *Impact of Climatic Variations on Storage Reservoir Systems*. Working Paper WP-90-20, IIASA, Laxenburg, Austria.

Kaczmarek, Z. (1990b) *On the Sensitivity of Runoff to Climate Change*. Working Paper WP-90-58, IIASA, Laxenburg, Austria.

Kaczmarek, Z. and Krasuski, D. (1991) *Sensitivity of Water Balance to Climate Change and Variability*. Working Paper WP-91-047, IIASA, Laxenburg, Austria.

Kaczmarek, Z., Strzepek, K.M., Somlyódy, L. and Priazhinskaya, V. (1996) *Water Resources Management in the Face of Climatic/Hydrologic Uncertainties*. Water Science and Technology Library 18. Kluwer, Dordrecht, The Netherlands.

Kahya, E. and Dracup, J.A. (1993) U.S. streamflow patterns in relation to El Niño/Southern Oscillation. *Water Resources Res.* **29**(8), 2491–503.

Kalinin, G.P., Markov, K.K. and Suetova, I.A. (1966) Fluctuations in the Earth's water bodies in the late geological past. *Oceanology* **VI**, 5–6.

Kalkstein, L. (ed.) (1991) *Global Comparisons of Selected GCM Control Runs and Observed Climatic Data*. US Environmental Protection Agency, Washington, DC, USA.

Kane, D.L., Hinzman, L.D., Woo, M.-K. and Everett, K.R. (1992) Arctic hydrology and climate change. In F.S. Chapin III, R.L.

Jeffries, J.F. Reynolds, G.R. Shaver, J. Svoboda and E.W. Chu (Eds.) *Arctic Ecosystems in a Changing Climate: An Ecophysiological Perspective*. Academic Press, San Diego, CA, pp. 35–57.

Karl, T.R. and Riebsame, W.E. (1989) The impact of decadal fluctuations in mean precipitation and temperature on runoff: A sensitivity study over the United States. *Climatic Change* **15**, 423–47.

Katz, R.W. (1996) Use of conditional stochastic models to generate climate change scenarios. *Climatic Change* **32**, 237–55.

Katz, R.W. and Brown, B.G. (1992) Extreme events in a changing climate: Variability is more important than averages. *Climatic Change* **21**, 289–302.

Kayano, M.T. and Moura, A.D. (1983) El Niño de 1982–83 e a precipitação sobre a América do Sul. *Rev. Bras. Geofísica*.

Kimball, B.A., La Morte, P.J., Pinter, L., Wale, G.W. and Garcia, R.L. (1993) Effects of free air CO_2 enrichment (FACE) on the energy balance and evapotranspiration. *Annual Meeting of the American Agronomy Society*, December 1993.

Kirshen, P.H. and Feunessey, N.M. (1992) *Potential Impacts of Climate Change upon the Water Supply of the Boston Metropolitan Area*. Draft Report to US Environmental Protection Agency.

Klemeš, V. (1983) Conceptualization and scale in hydrology. *J. Hydrology* **65**, 1–23.

Klemeš, V. (1985) *Sensitivity of Water Resource Systems to Climate Variations*. WCP No. 98, WMO, Geneva.

Klemeš, V. (1986) Operational testing of hydrological simulation models. *Hydrological Sci.* **31**, 13–24.

Klemeš, V. (1992) Implications of possible climate change for water management and development. *Water News: Canadian Water Research Association Newsletter*, April, pp. 2–3.

Klemeš, V. and Němec, J. (1983) Assessing the impacts of climate change on the development of surface water resources. In *Proceedings of the 2nd International Meeting on Statistical Climate, Lisbon*, 8.21–8.28.

Klige, R.K. (1985) *Changes in Global Water Exchange*. Nauka, Moscow.

Klige, R.K. (1994) *Prognostic Estimates Level Fluctuations in the Caspian Sea* (Meliatatsia i vodnoje khosiaistvo), No. 1, pp. 10–11 (in Russian)

Kopaliani, Z.D., Shiklomanov, I.A. and Georgievsky, V.Y. (1995) *Hydrological Assessment of the Water Balance of the Caspian Sea, Including Establishment of Hydrological Databanks and an Interrelated System for Monitoring and Transmission of Sea-level Fluctuations*. State Hydrological Institute, St. Petersburg.

Kousky, V.E., Kayano, M.T. and Cavalcanti, I.F.A. (1984) A review of the Southern Oscillation: Oceanic, atmospheric circulation changes and related rainfall anomalies. *Tellus* **A 36**, 490–504.

Krasovskaia, I. and Gottschalk, L. (1992) Stability of river flow regimes. *Nordic Hydrology* **23**, 137–54.

Krasovskaia, I., Gottschalk, L. and Arnell, N.W. (1994) Flow regimes in northern and western Europe: Development and application of procedures for classifying flow regimes. In *FRIEND: Flow Regimes from International Experimental and Network Data*. IAHS Publ. 221, pp. 185–92.

Krauss, T. (1996) Mackenzie GEWEX study science and planning workshop. *GEWEX News* **6**(1), 16.

Kuhl, S.C. and Miller, J.R. (1992) Seasonal river runoff calculated from a global atmospheric model. *Water Resources Res.* **28**(8), 2029–39.

Kwadijk, J. (1993) *The Impact of Climate Change on the Discharge of the River Rhine*. Netherlands Geographical Studies 171, University of Utrecht.

Kwadijk, J. and Middelkoop, H. (1994) Estimation of impact of climate change on the peak discharge probability of the River Rhine. *Climatic Change* **27**, 199–224.

Kwadijk, J. and Rotmans, J. (1995) The impact of climate change on the River Rhine: A scenario study. *Climatic Change* **30**, 397–425.

Lachenbruch, A.H. and Marshall, B.V. (1986) Changing climate: Geothermal evidence from permafrost in the Alaskan Arctic. *Science* **234**, 689–96.

Lamb, P.J. (1978) Large-scale tropical Atlantic circulation patterns associated with Saharan weather anomalies. *Tellus* **30**, 240–51.

Lamb, P.J. (1985) Rainfall in Subsaharan West Africa during 1941–83. *Z. Gletscherkunde und Glazialgeologie* **21**, 131–9.

Langbein, W.B. *et al.* (1949) *Annual Runoff in the United States*. Geological Survey Circular 52, US Department of the Interior, Washington, DC.

Lau, W.K.-M., Sud, Y.C. and Kim, J.-H. (1995) Intercomparison of hydrological processes in global climate models. In W.L. Gates (Ed.) *Proceedings of the 1st International AMIP Scientific Conference*, WCRP-92, WMO/TD-No. 732, pp. 71–76.

Laval, K. and Picon, L. (1986) Effect of a change of surface albedo of the Sahel on climate. *J. Atmos. Sci.* **4**, 2418–29.

Lawford, R.G. (1992) Science plan summary for the Canadian component of the first phase of the GEWEX programme. *GEWEX News*, pp. 5–7.

LBA Science Planning Group (1996) *Concise Experimental Plan, Large-scale Biosphere–Atmosphere Experiment in Amazonia*. Winand Staring Centre for Integrated Land, Soil and Water Research (SC-DLO), Wageningen, The Netherlands.

Lean, J., Bunton, C.B., Nobre, C.A. and Rowntree, P.R. (1996) The simulated impact of Amazonian deforestation on climate using measured ABRACOS vegetation characteristics. In J.H.C. Gash, C.A. Nobre, J.M. Roberts and R.L. Victoria (Eds.) *Amazonian Deforestation and Climate*. Chichester, Wiley, pp. 549–76.

Lean, J. and Rowntree, P.R. (1993) A GCM simulation of the impact of Amazonian deforestation on climate using an improved canopy representation. *Q. J. R. Meteorol. Soc.* **119**, 509–30.

Lean, J. and Warrilow, D.A. (1989) Simulation of regional climate impact of Amazon deforestation. *Nature* **342**, 411–13.

Leavesley, G.H., Branson, M.D. and Hay, L.E. (1992) Using coupled atmospheric and hydrologic models to investigate the effects of climate change in mountainous regions. In R. Herrmann (Ed.) *Managing Water Resources during Global Change*. Conf. Proc., American Water Resources Association, Bethesda, MD.

Lebel, T., Sauvageot, H., Hoepffner, M., Desbois, M., Guillot, B. and Hubert, P. (1992) Rainfall estimation in the Sahel: The EPSAT-NIGER experiment. *Hydrol. Sci. J.* **37**(3), 201–15.

Lebel, T., Taupin, J.D. and Le Barbé, L. (1995) Space–time fluctuations of rainfall during HAPEX–Sahel, *J. Hydrology*.

Leese, J.A. (1995) GCIP completes buildup for enhanced observing period. *GEWEX News* **5**(3), 3–5.

Leichenko, R.M. (1993) Climate change and water resource availability: An impact assessment for Bombay and Madras, India. *Water International* **18**(3), 147–56.

Lemmelä, R., Liebscher, H. and Nobilis, F., rapporteurs (1990) *Studies and Models for Evaluating the Impact of Climate Variability and Change on Water Resources*. WMO Regional Association VI (Europe), Working Group on Hydrology. Government Printing Centre, Helsinki.

Lemoalle, J. (1989) *Le Fonctionnement Hydrologique du lac Tchad au Cours d'une Période de Sécheresse (1973–1989)*. ORSTOM, Paris.

Lettenmaier, D.P. and Gan, T.Y. (1990) Hydrologic sensitivities of the Sacramento–San Joaquin River Basin of California to global warming. *Water Resources Res.* **26**(1), 69–86.

Lettenmaier, D.P. and Sheer, D.P. (1991) Climatic sensitivity of California water resources. *ASCE J. Water Resources Planning and Management* **117**(1), 108–25.

Leung, L.R., Wigmosta, M.S., Ghan, S.J., Epstein, D.J. and Vail, L.W. (1996) Application of a subgrid orographic precipitation/surface hydrology scheme to a mountain watershed. *J. Geophys. Res.*

Lins, H., Shiklomanov, I.A. and Stakhiv, E.Z. (1991) Impacts on hydrology and water resources. In J. Jaeger and H. Ferguson (Eds.) *Climate Change: Science, Impacts and Policy*. Proc. 2nd World Climate Conference, pp. 87–97.

Liu, C.Z. *et al.* (1995) *Study of the Impacts of Global Warming on Water Resources in China: Summary of Project Results*.

Love, P.K. and Henderson-Sellers, A. (1994) *Land Surface Climatologies of AMIP-PILPS Models and Identification of Regions for Investigation*. PCMDI Report.

Mabbutt, J.A. (1989) Impacts of carbon dioxide warming on climate and man in the semi-arid tropics. *Climatic Change* **15**, 191–221.

Machenhauer, B., Botzet, M., Jacob, D. and Christensen, J.H. (1996) Regionalization over Europe of global model simulations using the HIRHAM model. In S.J. Ghan, W.T. Pennell, K.L. Peterson, E. Rykiel, M.J. Scott, and L.W. Vail (Eds.) *Regional Impact of Global*

Climate Change: Assessing Change and Response at Scales that Matter. Battelle Press, Columbus, OH, pp. 23–50.

Mahé, G. (1993) Les écoulements fluviaux sur la façade atlantique de l'Afrique. Etude des éléments du bilan hydrique et variations inter-annuelles. Analyse de situations hydroclimatiques moyennes et extrêmes. Coll. Etudes et Theses, ORSTOM, Paris.

Maley, J. (1981) Etudes palynologiques dans le bassin du lac Tchad et paléoclimatologie de l'Afrique nord tropicale de 30000 ans BP à l'époque actuelle. Trav. et Doc. 129, ORSTOM, Paris.

Manabe, S. (1969) Climate and ocean circulation, 1: The atmospheric circulation and the hydrology of the earth's surface, *Monthly Weather Rev.* **97**, 739–74.

Manabe, S. and Stouffer, R.J. (1980) Sensitivity of a global climate model to increase of CO_2 concentration in the atmosphere. *J. Geophys. Res.* **85**, 5529–54.

Manabe, S. and Weatherald, R.T. (1987) Large scale changes in soil wetness induced by an increase in carbon dioxide. *J. Atmos. Sci.* **44**, 1211–35.

Margat, J. (1991) *Ressources en eau des pays africains: Utilisation et problèmes,* 7è Congrès Mondial des Ressources en Eaux, May 1991, Rabat, Morocco.

Mayo, L.R. and Trabant, D.C. (1984) Observed and predicted effects of climate change on Wolverine Glacier, southern Alaska. In J.H. McBeath (Ed.) *The Potential Effects of Carbon Dioxide-Induced Climate Changes in Alaska.* School of Agriculture and Land Resources Management, University of Alaska-Fairbanks, Misc. Pub. 83-1, pp. 114–23.

McCabe, G.J., Jr and Dettinger, M.D. (1995) Relations between winter precipitation and atmospheric circulation simulated by the Geophysical Fluid Dynamics Laboratory general circulation model. *Int. J. Climatology* **15**, 625–38.

McCabe, G.J., Jr and Hay, L.E. (1994) Hydrologic effects of hypothetical climate change in the East River basin, Colorado. *Water Resources Bull.*

McCabe, G.J., Jr. and Legates, D.R., (1995) Relationships between 700 h Pa height anomalies and April snowpack accumulations in the western USA. *Int. J. Climatology* **15**, 517–30.

McCabe, G.J., Jr and Wolock, D.M. (1992a) Effects of climatic change and climatic variability on the Thornthwaite moisture index in the Delaware River basin. *Climatic Change* **20**, 143–53.

McCabe, G.J., Jr and Wolock, D.M. (1992b) Sensitivity of irrigation demand in a humid temperate region to hypothetical climate change. *Water Resources Bull.* **28**, 535–43.

McCabe, G.J., Jr, Wolock, D.M., Tasker, G.D. and Ayers, M.A. (1991) Uncertainty in climate change and drought. In *Hydraulic Engineering 1991.* EE, IR, WW Div/ASCE, Nashville, TN.

McKnight, D.M. and Weiler, C.S. (1995) *Regional Assessment of Freshwater Ecosystems and Climate Change in North America.* US Government Printing Office, 1995-673-211/00033 Region No. 8.

Mekong Secretariat (1990) *Study of the Impacts of Climate Change on Water Resources in the Lower Mekong Basin.* Prepared by University of Colorado, Boulder, for the US EPA.

Miller, S.R. and Russell, G.C. (1992) The impact of global warming on river runoff. *J. Geophys. Res.* **93**(D3), 2757–64.

Mimikou, M. and Kouvopolous, Y.S. (1991) Regional climate change impacts: 1. Impacts on water resources. *Hydrol. Sci. J.* **36**, 247–58.

Mimikou, M., Kouvopoulos, Y., Cavadias, G. and Vayianos, N. (1991) Regional hydrological effects of climate change. *J. Hydrology* **123**, 119–46.

Minnis, P., Harrison, E.F., Stowe, L.L., Gibson, G.G., Denn, F.M., Doelling, D.R., Smith W.L. Jr (1993) Radiative climate forcing by the Mount Pinatubo eruption. *Science* **259**, 1411–15.

Mitchell, J.F.B. (1989) The 'greenhouse' effect and climate change. *Rev. Geophys.* **27**(1), 115–39.

Mitchell, J.F.B., Johns, T.C., Gregory, J.M. and Taft, S.F.B. (1995) Cimate response to increasing levels of greenhouse gases and sulphate aerosols. *Nature* **376**, 501–4.

Molion, L.C.B. (1976) *A Climatological Study of the Energy and Moisture Fluxes of the Amazonas Basin with Considerations of Deforestation Effects.* INPE 923-TPT/035, S.J. Campos, SP.

Molion, L.C.B. (1987) Micrometeorology of an Amazonian rainforest. In R.E. Dickinson (Ed.) *The Geophysiology of Amazonia.* Wiley, New York, pp. 255–72.

Molion, L.C.B. (1990) Climate variability and its effects on Amazonian hydrology. *Interciencia* **15**(6), 367–72.

Molion, L.C.B. (1993) Amazonian rainfall and its variability. In M. Bonell, J. Gladwell and M. Hufschmidt (Eds.) *Hydrology and Water Management in the Humid Tropics.* Cambridge University Press/UNESCO, pp. 99–111.

Molion, L.C.B. (1994) Efeitos de vulcões no clima. *Cadernos de Geociências* **12**, 13–23, IBGE, Rio de Janeiro.

Molion, L.C.B. (1995) Global warming: A critical review. *Interciencia.*

Molion, L.C.B. and Moraes, J.C. (1987) Oscilação Sul e descarga de rios na América do Sul tropical. *Rev. Bras. Eng. Caderno de Hidrologia* **5**(1), 53–63.

Mota, R. and Tucci, C. (1984) *Modelo IPH III, Rev. Bras. Eng.* RBE/CRH, ABRH, Sao Paulo, Brazil.

Moore, R.J. (1985) The probability-distribution principle and runoff production at point and basin scales. *Hydrol. Sci. J.* **30**, 263–97.

Murphy, J.M. and Mitchell, J.F.B. (1995) Transient response of the Hadley Centre coupled ocean–atmosphere model to increasing carbon dioxide. Part II: Spatial and temporal structure of response. *J. Climate* **8**, 57–80.

Nachtigall, G. (1881) *Sahara et Soudan.* Hachette, Paris.

NASA (1991) *EOS Reference Handbook.* National Aeronautics and Space Administration, Goddard Space Flight Center, Greenbelt, MD.

Nash, L.L. and Gleick, P.H. (1991) Sensitivity of streamflow in the Colorado Basin to climate changes. *J. Hydrology* **125**, 221–41.

Nash, L.L. and Gleick, P.H. (1993) *The Colorado River Basin and Climatic Change: The Sensitivity of Streamflow and Water Supply to Variations in Temperature and Precipitation.* EPA 230-R-93-009, U.S. EPA, Office of Policy, Planning and Evaluation, Washington, DC.

Nathan, R.J., McMahon, T.A. and Finlayson, B.L. (1988) The impact of the greenhouse effect on catchment hydrology and storage–yield relationships in both winter and summer rainfall zones. In G.I. Pearman (Ed.) *Greenhouse: Planning for Climate Change.* Div. of Atmospheric Research, CSIRO, East Melbourne, Australia.

Němec, J. and Schaake, J. (1982) Sensitivity of water resource systems to climate variation. *Hydrol. Sci. J.* **3**: 327–43.

Ng, H.Y.F. and Marsalek, J. (1992) Sensitivity of streamflow simulation to changes in climatic inputs. *Nordic Hydrology* **23**, 257–72.

Nicholson, S.E. (1980) Saharan climates in historic times. In Williams and H. Faure (Eds.) *The Sahara and the Nile.* Balkema, Rotterdam, pp. 173–200.

Nicholson, S.E. (1981a) The historical climatology of Africa. In T.M.L. Wigley *et al.* (Eds.) *Climate and History.* Cambridge University Press, Cambridge, pp. 249–70.

Nicholson, S.E. (1981b) Rainfall and atmospheric circulation during drought periods and wetter years in West Africa. *Monthly Weather Rev.* **109**, 2191–208.

Nicholson, S.E., Jeeyoung, K. and Hoopingarner, J. (1988) *Atlas of African Rainfall and its Interannual Variability.* Dept. of Meteorology, Florida State University, Tallahassee, FL.

Nobre, C. (1993) GCM model studies. In P. Sellers, C. Nobre, D. Fitzjarrald, P. Try and D. Lucid (Eds.) *A Preliminary Science Plan for a Large Scale Biosphere–Atmosphere Field Experiment in the Amazon Basin.* ISLSCP/GEWEX, Washington, DC.

Nobre, C., Fitzjarrald, D. and Sellers, P. (1993) Conceptual plan for LAMBADA BATERISTA. In P. Sellers, C. Nobre, D. Fitzjarrald, P. Try and D. Lucid (Eds.) *A Preliminary Science Plan for a Large Scale Biosphere–Atmosphere Field Experiment in the Amazon Basin.* ISLSCP/GEWEX, Washington, DC.

Nobre, C., Sellers, P.J. and Shukla, J. (1991) Amazonian deforestation and regional climate change. *J. Climatology* **10**(4), 957–88.

Nobre, C. and Shuttleworth, J. (1993) Anglo–Brazilian Amazonian Climate Observational Study (ABRACOS). In P. Sellers, C. Nobre, D. Fitzjarrald, P. Try and D. Lucid (Eds.) *A Preliminary Science Plan for a Large Scale Biosphere–Atmosphere Field Experiment in the Amazon Basin.* ISLSCP/GEWEX, Washington, DC.

Nophadol, L. and Hemanth, E.J. (1992) In: R. Herrmann (Ed.) *Impact Assessment of Global Warming on Rainfall–Runoff Characteristics in a*

Tropical Region (Sri Lanka). AWRA Symp. on Managing Water Resources During Global Change, pp. 547–56.

Oldekop, E. (1911) *Evaporation from River Basin Surfaces*. Yuriev.

Olivry, J.-C., Bricquet, J-P. and Mahé, G. (1993) *Vers un Appauvrissement Durable des Ressources en Eau de l'Afrique Humide?* IAHS Publ. No. 216, Yokohama, pp. 67–78.

Olivry, J.-C. and Chastanet, M. (1986) *Evolution du Climat dans le Bassin du Fleuve Sénégal (Bakel) Depuis le Milieu du 19ème Siècle*. Coll. Trav. et Doc. No. 197, ORSTOM, Paris, pp. 337–43.

Ozga-Zielinska, M., Brzezinski, J. and Feluch, W. (1994) *Meso-scale Hydrologic Modelling for Climate Impact Assessments: A Conceptual and a Regression Approach*. Collaborative Report CP-94-10, IIASA, Laxenburg, Austria.

Palmer, T.N. (1986) Influence of the Atlantic, Pacific and Indian Oceans on Sahel rainfall. *Nature* **320**, 251–3.

Palutikof, J. (1987) *Some Possible Impacts of Greenhouse Gas Induced Climatic Change on Water Resources in England and Wales: The Influence of Climate Change and Climatic Variability on the Hydrologic Regime and Water Resources*. IAHS Publ. 168, pp. 585–96.

Panagoulia, D. (1992) Impacts of GISS-modelled climate changes on catchment hydrology. *Hydrological Sci.* **37**(2), 141–63.

Parry, M.L., Blantran de Rozari, Chong, A.L. and Panich, S. (Eds.) (1991) *The Potential Socio-Economic Effects of Climate Change in South-East Asia*. UNEP, p. 123.

Piper, B.S., Plinston, D.T. and Sutcliffe, J.V. (1986) The water balance of Lake Victoria. *Hydrol. Sci. J.* **31**(1), 25–38.

Poiani, K.A. and Johnson, W.C. (1993) Potential effects of climate change on a semi-permanent prairie wetland. *Climatic Change* **24**, 213–32.

Popper, W. (1951) *The Cairo Nilometer*. University of California Press, Berkeley.

Porter, J.W. and McMahon, T.A. (1971) A model for the simulation of streamflow data from climatic records. *J. Hydrology* **13**, 297–324.

Pouyaud, B. (1987) *Variabilité Spatiale et Temporelle des Bilans Hydriques de Quelques Bassins Versants d'Afrique de l'Ouest en Liaison avec les Changements Climatiques*. IAHS Publ. No. 168, pp. 447–61.

Pouyaud, B. and Colombani, J. (1989) Les variations extrêmes du lac Tchad: l'assèchement est-il possible? *Ann. Géographie, Paris* **98**(545), 1–23.

Quinn, W.H. and Neal, V.T. (1987) El Niño occurrences over the past four and a half centuries. *J. Geophys. Res.* **92**(C13), 14 449–14 461.

Randall, D.A. and Gleckler, P.J. (1995) Diagnosis of simulated ocean surface heat fluxes and the implied partitioning of meridional heat transport between the atmosphere and the ocean. In W.L. Gates (Ed.) *Proceedings 1st International AMIP Scientific Conference*, WCRP-92, WMO/TD No. 732, WMO, Geneva, pp. 25–30.

Rao, V.B. and Hada, K. (1990) Characteristics of rainfall over Brazil: Annual variations and Southern Oscillation, *Theoretical and Applied Climatology*.

Redmond, K.T. and Koch, R.W. (1991) Surface climate and streamflow variability in the western United States and their relationship to large-scale circulation indices. *Water Resources Res.* **27**(9), 2381–99.

Revelle, R.R. and Waggoner, P.E. (1983) Effects of a carbon dioxide-induced change on water supply in the Western United States. In *Changing Climate*. National Academy of Sciences, National Academy Press, Washington, DC, pp. 419–32.

Roald, L.A., Nordseth, K. and Hassel, K.A. (Eds.) (1989) *FRIENDs in Hydrology*. IAHS Publ. 187.

Roback, A.C., Schlosser, C.A. Vinnikov, K.Y. and Liu, S. (1995) Validation of humidity, moisture fluxes, and soil moisture in GCMS: Report of AMIP diagnostic subproject 11 Part 1: Soil moisture. In W.L. Gates (Ed.) *Proceedings 1st International AMIP Scientific Conference*, WCRP-92, WMO/TD No. 732, WMO, Geneva, pp. 1–8.

Robinson, P.J. and Finkelstein, P.L. (1989) *Strategies for the Development of Climate Scenarios for Impact Assessment*. U.S. Environmental Protection Agency, Research Triangle Park, NC, Phase 1 Final Report.

Rognon, P. (1989) *Biographie d'un Désert*, Coll. Scientifique SYNTHESE, Plon, Paris.

Rohlfs, G. (1874) *Quer durch Africa: Reise vom Mittelmeer nach dem Tschadsee und zum Golf von Guinea*, 2 vols. Leipzig.

Roots, E.F. (1989) Climate change: High-latitude regions. *Climatic Change* **15**, 223–53.

Roset, J.-P. (1987) *Néolitisation, Néolithique et Post-neolithique au Niger Oriental*. Congrès INQUA, Ottowa, pp. 203–14.

Rowell, D.P. and Blondin, C. (1990) The influence of soil wetness distribution on short range rainfall forecasting in the west African Sahel. *Q. J. R. Meteorol. Soc.* **116**, 1471–85.

Rozari, M.B. *et al.* (1990) *Socioeconomic Impacts of Climate Change: Indonesian Report*. Report submitted to UNEP.

Running, S.W. and Coughlan, J.C. (1988) A general model of forest ecosystem processes for regional applications. I: Hydrologic balance, canopy gas exchange, and primary production processes. *Ecological Modelling* **42**, 125–54.

Running, S.W. and Nemani, R.R. (1991) Regional hydrologic carbon balance responses of forests resulting from potential climate change. *Climatic Change* **19**, 349–68.

Sælthun, N.R., Bogen, J., Flood, M.H. *et al.* (1990) *Climate Change Impact on Norwegian Water Resources*. Norwegian Water Resources and Energy Administration Publ. 42.

Saïd, R. (1994) Origin and evolution of the River Nile. In P.P. Howell and J.A. Allan (Eds.) *The Nile: Sharing a Scarce Resource*. Cambridge University Press, Cambridge, pp. 17–26.

Salati, E., Marques, J. and Molion, L.C.B. (1978) Origen e distribuição das chuvas na Amazonia. *Interciência* **3**, 200–6.

Saleh, M., Strzepec, K. and Yates, D. (1994) Potential climate change impacts on the Nile Basin. *Proceedings 8th IWRA World Congress on Water Resources,* Cairo, November 1994.

Salewicz, K. (1995) Impact of climate change on the Lake Kariba hydropower scheme. In *Water Resources Management in the Face of Climatic and Hydrologic Uncertainties*. Kluwer, Dordrecht.

Salinger, M. and Hicks, D. (1989) *Regional Climate Change Scenarios*. New Zealand Ministry for the Environment.

Sanderson, M. and Wong, L. (1987) Climate change and Great Lakes water levels. In S. Solomon *et al.* (Eds.) *The Influence of Climatic Change and Climate Variability on Hydrological Regimes and Water Resources*. IAHS Publ. No. 168, pp. 441–87.

Santer, B.D. *et al.* (1990) *Developing Climate Scenarios from Equilibrium GCM Results*. Report No. 47, Max Planck Institut für Meteorologie, Hamburg.

Sato, T. *et al.* (1989) Effects of implementing the Simple Biospheric Model (SIB) in a general circulation model. *J. Atmospheric Sci.* **46**, 2757–82.

Schaake, J.C. (1990) From climate to flow. In P.E. Waggoner (Ed.) *Climate Change and U.S. Water Resources*. Wiley, New York, pp. 177–206.

Schlesinger, M.E. and Zhao, Z. (1987) *Seasonal Climate Changes Induced by Doubled CO_2 as Simulated by the OSU Atmospheric OCM/Mixed Layer Model*. Report 70, Oregon State University Climate Institute, Corvallis, OR.

Sellers, P.J. and Hall, F.G. (1992) FIFE in 1992: Results, scientific gains and future research directions. *J. Geophys. Res.* **97**(D17), 19 091–109.

Sellers, P.J. and Hall, F.G. (1994) Boreal forest and climate change. *GEWEX News* **4**(4), 6–7.

Sellers, P.J., Hall, F.G., Asrar, G., Strebel, D.E. and Murphy, R.E. (1992) An overview of the First International Satellite Land Surface Climatology Project (ISLSCP) Field Experiment (FIFE). *J. Geophys. Res.* **97**(D17), 18 345–371.

Sellers, P.J. *et al.* (1989) Calibrating the Simple Biospheric Model (SIB) for Amazonian tropical forest using field and remote sensing data, Part I: Average calibration with field data. *J. Applied Meteorol.* **28**(8), 727–59.

Servant-Vildary, S. (1978) *Etude des diatomées et paléolimnologie du bassin du Tchad au Cenozoïque Supérieur*. Coll. Trav. et Doc. No. 84, ORSTOM, Paris.

Seuna, P., Gustard, A., Arnell, N.W. and Cole, G.A. (Eds.) (1994) *FRIEND: Flow Regimes from International Experimental and Network Data*. IAHS Publ. No. 221.

Shiklomanov, I.A. (1976) *Hydrological Aspects of the Caspian Sea Problem*. Hydrometeoizdat, Leningrad (in Russian).

Shiklomanov, I.A. (1988) *Investigation of Land and Water Resources: Conclusions, Problems, Perspective*. Hydrometeoizdat, Leningrad (in Russian).

Shiklomanov, I. (1989a) Anthropogenic climate change, water resources, and water management problems. In *Proceedings of a Conference on Climate and Water*. Helsinki, Finland, WMO, Geneva.

Shiklomanov, I. (1989b) *Man's Impact on River Runoff*. Hydrometeoizdat, Leningrad (in Russian).

Shiklomanov, A.I. (1994) On the effect of the anthropogenic changes in global climate on runoff in the Yenisey basin. *Meteorology and Hydrology*, No. 2, 84–93 (in Russian).

Shiklomanov, I.A. and Babkin, V.I. (1992) Climate change and water management. *Meteorology and Hydrology*, No. 8, 38–43 (in Russian).

Shiklomanov, I.A. and Lins, H. (1991) Influence of climate change on hydrology and water management. *Meteorology and Hydrology*, No. 4, 51–66 (in Russian).

Shiklomanov, I.A., Lins, H. and Stakhiv, E. (1990) Hydrology and water resources. In W. Tegart, G. Sheldon and D. Griffiths (Eds.) *The IPCC Impact Assessment*.

Shiklomanov, I.A. and Markova, O.L. (1987) *Global Problems of Water Availability and Water Transfers*. Leningrad, Hydrometeoizdat (in Russian).

Shilo, N.A. and Krivoshei, M.I. (1989) Interrelation between water level fluctuations in the Caspian Sea and tensions in the Earth's crust. *Vestnik AN SSSR*, No. 6, 83–90.

Shukla, J. Nobre, C. and Sellers, P. (1990) Amazonian deforestation and climate change. *Science* 247, 1322–5.

Shuttleworth, W.J. (1987) International investigations on large-scale evaporation. In *Evaporation and Weather*, Proceedings and Information No. 39, TNO Committee on Hydrological Research, The Hague, The Netherlands, pp. 85–93.

Shuttleworth, W.J. (1989) Micrometeorology of temperate and tropical forest. *Phil. Trans. R. Soc. London, B* 324, 299–334.

Simpson, J., Adler, R.F. and North, G.R. (1988) A proposed Tropical Measuring Mission (TRMM) satellite. *Bull. Am. Meteorol. Soc.* 69, 278–95.

Sircoulon, J. (1976) Les données hydropluviométriques de la sécheresse récente en Afrique intertropicale. Comparaison avec les sécheresses 1913 et 1940. Cah. ORSTOM, sér. Hydrology XIII(2), 75, 174.

Sircoulon, J. (1987) *Variation des Débits des Cours d'Eau et des Niveaux des Lacs en Afrique de l'Ouest Depuis le Début du 20ème Siècle*. IAHS Publ. No. 168, pp. 13–25.

Sircoulon, J. (1990) *Impact possible des changements climatiques à venir sur les ressources en eau des régions arides et semi-arides. Comportement des cours d'eau tropicaux, des rivières et des lacs en zone sahélienne*. WCAP-12; WMO/TD. No. 380, WMO, Geneva.

Skiles, J.W. and Hanson, J.D. (1994) Responses of arid and semiarid watersheds to increasing carbon dioxide and climate change as shown by simulation studies. *Climatic Change* 26, 377–97.

Solomon, S.I., Beran, M. and Hogg, W. (1987) *The Influence of Climate Change and Variability on Hydrologic Regimes and Water Resources*. Proc. Vancouver Symposium, August 1987, IAHS Publ. No. 168.

Sorooshian, S. and Gupta, V.K. (1983) Automatic calibration of conceptual rainfall–runoff models: The question of parameter observability and uniqueness. *Water Resources Res.* 19(1), 260–8.

Srinivasan, G., Holme, M., Jones, C.G., Jones, P.D. and Osborn, T.J. (1995) An evaluation of the spatial and interannual variability of tropical precipitation as simulated by GCMS. In W.L. Gates (Ed.) *Proceedings 1st International AMIP Scientific Conference*, WCRP-92, WMO/TD No. 732, WMO, Geneva, pp. 193–8.

Stakhiv, E., Lins, H. and Shiklomanov, I. (1992) Hydrology and water resources. In W. Tegart and G. Sheldon (Eds.) *Supplementary Report to the IPCC Impact Assessment*. Australian Government Publ. Ser.

Stakhiv, E., Shiklomanov, I. and Lins, H. (1993) *Hydrology and Water Resources*. IPCC Report.

Stockle, C.O., Dyke, P.T., Williams, J.R., Jones, C.A. and Rosenberg, N.J. (1992b) Estimation of the effects of CO_2-induced climate change on growth and yield of crops. II: Assessing the impacts on maize, wheat, and soybean in the midwestern USA. *Agric. Systems.* 38, 239–56.

Stockle, C.O., Williams J.R., Rosenberg, N.J. and Jones, C.A. (1992a) Estimation of the effects of CO_2-induced climate change on growth and yield of crops. I: Modification of the EPIC model for climate change analysis. *Agric. Systems* 38, pp. 225–38.

Stockton, C.W. and Bogess, W.R. (1979) *Geohydrological Implications of Climate Change on Water Resources Development*. US Army Coastal Engineering Research Center, Fort Belvoir, VA.

Stouffer, R.J., Manabe, S. and Bryan, K. (1989) Interhemispheric asymmetry in climate response to a gradual increase of atmospheric CO_2. *Nature* 342, 660–2.

Street, F.A. and Grove, A.T. (1979) Global maps of lake level fluctuations since 30000 BP. *Quaternary Res.* 12, 83–118.

Sud, Y.C. and Molod, A. (1988) A GCM simulation study of the influence of Saharan evapotranspiration and surface albedo anomalies on July circulation and rainfall. *Monthly Weather Rev.* 116, 2388–400.

Sutcliffe, J.V. and Knott, D.G. (1987) *Historical Variations in African Water Resources*. IAHS Publ. No. 168, pp. 463–75.

Sutcliffe, J.V. and Lazenby, J. (1994) Hydrological data requirements for planning Nile Management. In P.P. Howell and J.A. Allan (Eds.) *The Nile: Sharing a Scarce Resource*. Cambridge University Press, Cambridge, pp. 163–92.

Szilugyi, F. and Smolyódy, L. (1991) *Potential Impacts of Climate Change on Water Quality in Lakes*. IAHS Publ. No. 205, pp. 79–86.

Tallaksen, L. and Hassel, K.A. (Eds.) (1992) *Climate Change and Evaporation Modelling*. Seminar on evapotranspiration models for simulating climate change impact on the catchment water balance, Vettre, Norway, March 1992; Nordic Hydrologic Programme, NHP Rep. No. 31, Nordic Coordinating Committee for Hydrology (KOHYNO).

Thornthwaite, C.W. (1948) An approach toward a rational classification of climate. *Geog. Rev.* 38, 55–94.

Thornthwaite, C.W. and Mather, J.R. (1955) *The Water Balance*. Publications in Climatology, Drexel Institute of Technology, Laboratory of Climatology, VIII, p. 1.

Toth, F.L. (1993) *Policy Responses to Climate Change in Southeast Asia*. IIASA, Laxenburg, Austria, pp. 304–23.

Tucci, C.E.M. and Damiani, A. (1991) International studies on climate change impacts: Uruguay River Basin. *Proceedings of a Workshop on the Analysis of Potential Climate Change in the Uruguay River Basin*. Tech. Rep. No. 25, Institute of Hydraulic Research, Federal University of Rio Grande do Sul, Brazil.

Tucci, C. and Damiani, A. (1994) *Potencial Impacto da Modificação Climática no Rio Uruguai*. Revista Brasileira de Engenharia, RBE-CRH, vol. 12(2), Associação Brasileira de Recursos Hídricos, pp. 5–34.

U.S. Department of Commerce (1994) *The Great Flood of 1993*. NOAA Natural Disaster Survey Report, Washington, DC.

U.S. Environmental Protection Agency (1984) *User's Manual for Hydrological Simulation Program-FORTRAN (HSPF)*, EPA-600/3-84-066, Environmental Research Laboratory, Athens, GA.

Van Blarcum, S.C., Miller, J.R. and Russell, G.L. (1995) High-latitude river runoff in a doubled CO_2 climate. *Climatic Change* 30, 7–26.

Varushchenko, S.I., Varushchenko, A.N. and Klige, R.K. (1987) *Change in the Regime of the Caspian Sea and in Endorheic Water Bodies during Paleotime*. Nauka, Moscow.

Vehviläinen, B. and Lohvansuu, J. (1991) The effect of climate change on discharges and snow cover in Finland. *Hydrol. Sci. J.* 36, 109–21.

Verstraete, M.M. and Dickinson, R.E. (1986) Modeling surface processes in atmospheric general circulation models. *Ann. Geophys.* B4(4), 357–64.

Victoria, R., Mortatti, J., Richey, J., Dunne, T. and Zhang Z. (1993) Carbon in the Amazon River Basin (CAMREX). In P. Sellers, C. Nobre, D. Fitzjarrald, P. Try and D. Lucid (eds.) *A Preliminary Science Plan for a Large-Scale Biosphere Atmosphere Field Experiment in the Amazon Basin*. ISLSCP/GEWEX, Washington, DC.

Viner, D. and Hulme, M. (1993) *The UK Met. Office High-Resolution GCM Equilibrium Experiment (UKHI): Climate Impacts Link Technical Report 1*. Climate Research Unit, University of East Anglia.

Walsh, J.E., Meleshko, V., Tao, X. and Kattsov, V. (1995) AMIP model simulations of the polar regions. In W.L. Gates (Ed.) *Proceedings of the 1st International AMIP Scientific Conference*, WCRP-92, WMO/TD-No. 732, pp. 31–6.

Washington, W.M. and Meehl, G.A. (1984) A seasonal cycle experiment on the climate sensitivity due to a doubling of CO_2 with an atmospheric general circulation model coupled to a simple mixed layer ocean model. *J. Geophys. Res.* 89, 9475–503.

Washington, W.M. and Meehl, G.A. (1989) Climate sensitivity due to increased CO$_2$: Experiments with a coupled atmosphere and ocean general circulation model. *Climate Dynamics* **4**, 1–38.

Water Assessment (1993) *Joint ADB/IBRD/UNDP Subsaharan Africa Hydrological Assessment (Project RAF/87/030): West African Countries.* Prepared with Mott-MacDonald, BCEOM, SOGREAH and ORSTOM.

Waterstone, M. *et al.* (1995) *Future Water Resources Management in the Upper Rio Grande Basin.* Report for US Army Institute for Water Resources, Fort Belvoir, VA.

WCRP (1992) *Science Plan for the GEWEX Continental-Scale International Project (GCIP).* WCRP-67, WMO/TD No. 461, World Climate Research Programme, WMO, Geneva.

Weatherald, R.T. and Manabe, S. (1986) An investigation of cloud cover change in response to thermal forcing. *Climatic Change* **8**, 5–24.

Wigley, T.M.L and Jones, P.D. (1985) Influence of precipitation changes and direct CO$_2$ effects on streamflow. *Nature* **314**, 163.

Wilks, D.S. (1992) Adapting stochastic weather generation algorithms for climate change studies. *Climatic Change* **22**, 67–84.

Williams, J.R., Jones, C.A. and Dyke, P.T. (1984) A modeling approach to determine the relationship between erosion and soil productivity. *Trans. Am. Soc. Agricultural Engineers* **27**, 129–44.

Williamson, D.L., Kiehl, J.T., Ramanathan, V., Dickinson, R.E. and Hack, J.J. (1987) *Descriptions of NCAR Community Climate Model (CCM1).* Tech. Note NCAR/tn-285+str, National Center for Atmospheric Research, Boulder, CO.

Wilson, C.A. and Mitchell, J.F.B. (1987) A doubled CO$_2$ climate sensitivity experiment with a GCM including a simple ocean. *J. Geophys. Res.* **92**, 13 315–343.

Wilson, L.L., Lettenmaier, D.P. and Skyllingstad, E. (1992) A hierarchical stochastic model of large-scale atmospheric circulation patterns and multiple station daily precipitation. *J. Geophys. Res.* **97**(D3), 2791–809.

Winter, T.C. (1989) Distribution of the difference between precipitation and open-water evaporation in North America. In *The Geology of North America*, Vol. O-1. Geological Society of America, Boulder, CO (Plate 2).

WMO (1975) *Intercomparison of Conceptual Models Used in Operational Hydrological Forecasting.* Operational Hydrology Report No. 7, WMO No. 429, WMO, Geneva.

WMO (1979) *World Climate Conference.* Abstracts of reports, WMO, Geneva.

WMO (1985a) *Sensitivity of Water Resource Systems to Climate Variations.* WCP-98, WMO, Geneva.

WMO (1985b) *Intercomparison of Models of Snowmelt Runoff.* Operational Hydrology Report No. 23, WMO No. 646, WMO, Geneva.

WMO (1989a) *Proceedings of a Conference on Climate and Water.* Helsinki, Finland, 2 vols, WMO, Geneva.

WMO (1989b) *Statistics on Regional Networks of Climatological Stations* (based on the INFOCLIMA World Inventory). Vol. II: Region I, Africa, WCDP-7, WMO/TD No. 305, WMO, Geneva.

WMO (1990) *Studies and Models for Evaluating the Impacts of Climate Variability and Change on Water Resources.* WMO Regional Association VI (Europe), Helsinki, Finland WMO, Geneva.

WMO (1992a) *Science Plan for the GEWEX Continental-Scale International Project (GCIP).* WCRP-67, WMO/TD No. 461, WMO, Geneva.

WMO (1992b) *Scientific Concept of the Arctic Climate Study (ACSYS).* WCRP-72, WMO/TD No. 486, WMO, Geneva.

Woo, M.-K., Lewkowicz, A.G. and Rouse, W.R. (1992) Response of the Canadian permafrost environment to climate change. *Physical Geography* **134**, 287–317.

Wood, E.F., Sivipalan, M., Beven, K. and Band, L. (1988) Effects of spatial variability and scale with implications to hydrologic modeling. *J. Hydrology* **102**, 29–47.

Wood, E.F., Sivipalan, M. and Beven, K. (1990) Similarity and scale in catchment storm response. *Rev. Geophys.* **28**, 1–18.

Woolhiser, D.A. and Brakensiek, D.L. (1982) Hydrologic system synthesis. In C.T. Haan, H.P. Johnson and D.L. Brakensiek (Eds.) *Hydrologic Modeling of Small Watersheds.* ASAE Monograph No. 5, American Society of Agricultural Engineers.

Xue, Y. and Shukla, J. (1993) The influence of land surface properties on the Sahelian climate. Part I: Desertification, *J. Climate* **6**, 2232–45.

Yamada, T. (1989) *Tasks for River Administration in Relation to Global Environmental Issues.* Japan Ministry of Construction (River Bureau) (unpublished).

Yasunari, T. (1996) *The GEWEX Asian Monsoon Experiment (GAME).* Report to the 8th Session of GEWEX SSG, January 15–19, Irvine, CA.

Zaikov, A. (1946) Water balance of the Caspian Sea and reasons for the fall in water level. *Trudy NIH GUGMS*, Ser. 4, **88**, 19 (in Russian).

Zillman, J.W. (1989) Climate variability and change: Implications for the Murray–Darling basin. *12th International Symposium on the Murray–Darling Basin: A Resource to be Managed.* Australian Academy of Technology.

Zwerver, S., van Rompaey, R.S.A.R., Kok, M.T.J. and Berk, M.M. (Eds.) (1995) *Climate Change Research, Evaluation, and Policy Implications.* Proceedings of the International Climate Change Research Conference, Maastricht, The Netherlands, December 1994, Studies in Environmental Science, vols. 65A and 65B. Elsevier, Amsterdam.

Appendix: Acronyms and abbreviations

ABLE	Amazon Boundary Layer Experiment
ABRACOS	Anglo–Brazilian Amazonian Climate Observation Study
ACSYS	Arctic Climate System Study
AMBIACE	Amazon Ecology and Atmospheric Chemistry Experiment (LBA, Brazil)
AMHY	Alpine and Mediterranean HYdrology project
AMIP	Atmospheric Model Intercomparison Project (WCRP)
AP	Accelerated Policy
ARME	Amazon Region Micrometeorological Experiment
AVHRR	Atmospheric Very High Resolution Radiometer
BAHC	Biospheric Aspects of the Hydrologic Cycle (IGBP/WCRP core project)
BALTEX	BALTic EXperiment
BATERISTA	Biosphere–Atmosphere Transfers and Ecological Research In Situ Studies in Amazonia (LBA, Brazil)
BATS	Biosphere–Atmosphere Transfer Scheme
BaU	'Business-as-Usual' scenario
BGC	BioGeochemical Cycles Study
BOREAS	Boreal Ecosystem–Atmosphere Study
BRDF	Bidirectional Reflectance Distribution Function
CAMREX	Carbon in the AMazon River EXperiment (Brazil)
CCIRG	Climate Change Impacts Review Group (UK)
CCC	Canadian Climate Centre
CCM	Community Climate Model (NCAR)
CENA	Centro de Energia Nuclear na Agricultura (Brazil)
CFCs	Chlorofluorocarbons
CGER	Center for Global Environmental Research (Tsukuba, Japan)
CNRM	Centre National de Recherches Météorologiques
COARE	Coupled Ocean–Atmosphere Experiment
COHMAP	COoperative Holocene MApping Project

CPTEC	Centro de Previsao de Temp e Estudos Climaticos (Brazil)
CRSS	Colorado River System Simulation model (US Bureau of Reclamation)
CSA	Continental-Scale Area (GEWEX)
CSD	UN Commission on Sustainable Development
CSIRO	Commonwealth Scientific and Industrial Research Organization (Australia)
DNAEE	Brazilian National Department of Water and Power
ECHIVAL	European International Project on Climate and Hydrological Interactions between Vegetation, Atmosphere and Land Surfaces (part of EPOCH)
ECMWF	European Centre for Medium-range Weather Forecasting (Reading, UK)
EFEDA	ECHIVAL Field Experiment in Desertification-threatened Areas
ENSO	El Niño–Southern Oscillation
EOS	Earth Observing System
EODIS	EOS Data and Information System
EPA	Environmental Protection Agency (USA)
EPIC	Erosion Productivity Impact Calculator
EPOCH	European Programme on Climate and Natural Hazards
EPSAT-Niger	Estimation of Precipitation by SATellites in Niger
ERS	European Remote Sensing satellite
ESA	European Space Agency
ESCAPE	Evaluation of Strategies to Address Climate Change by Adapting to and Preventing Emissions (RIVM project)
ET	Evapotranspiration
FIFE	First International Satellite Land Surface Climatology Project (ISLSCP) Field Experiment
FLUAMAZ	Amazon Moisture Flux Experiment (Brazil)
FRIEND	Flow Regimes from International Experimental and Network Data

FOREST-BGC	Forest BioGeochemical Cycles model
GAIM	Global Analysis, Interpretation and Modelling project (IGBP)
GAME	GEWEX Asian Monsoon Experiment
GARP	Global Atmospheric Research Program
GATE	GARP Atlantic Tropical Experiment
GCIP	GEWEX Continental-scale International Project
GCOS	Global Climate Observing System
GCM	General Circulation Model
GCTE	Global Change and Terrestrial Ecosystems Project (IGBP)
GEMS	Global Environmental Monitoring System (Canada)
GEWEX	Global Energy and Water Cycle EXperiment
GFDL	Geophysical Fluid Dynamics Laboratory (Princeton, USA)
GIS	Geographical Information System
GISS	Goddard Institute for Space Sciences (USA)
GLUE	Generalized Likelihood Uncertainty Estimation
GOES	Geostationary Operational Environmental Satellite
GOOS	Global Ocean Observing System
GRACE	Groundwater Resources And Climate Change Effects project (EC)
GTOS	Global Terrestrial Observing System
HAPEX	Hydrologic-Atmospheric Pilot EXperiment
HAPEX-MOBILHY	Hydrologic-Atmospheric Pilot EXperiment-Modélisation du Bilan Hydrique
HAPEX-Niger	Hydrologic-Atmospheric Pilot EXperiment-Niger
HAPEX-Sahel	Hydrologic-Atmospheric Pilot EXperiment-Sahel
HCCP	Hadley Centre for Climate Prediction (UKMO)
HIRHAM	HIgh-Resolution local Area Model
HSPF	Hydrologic Simulation Program – FORTRAN model (US EPA)
HYTRECS	HYdrology of TRopical ECosystems (Brazil)
IAHS	International Association of Hydrological Sciences
IAMAP	International Association of Meterology and Atmospheric Physics of the IUGG
ICA	International Council of Archives
ICASVR	International Committee on Atmosphere, Soil and Vegetation Relations (IAHS)
ICSU	International Council of Scientific Unions (ICSU and WRCP)
IGBP	International Geosphere–Biosphere Programme (ICSU and WCRP)
IGPO	International GEWEX Project Office
IH	Institute of Hydrology, Wallingford, UK
IHP	International Hydrological Programme (UNESCO)
IIASA	International Institute for Applied Systems Analysis (Austria)
IMAGE	Integrated Model for the Assessment of the Greenhouse Effect (RIVM project)
INPA	Instituto Nacional de Pesquisas da Amazonia (Institute for Amazon Research, Brazil)
INPE	Instituto Nacional de Pesquisas Espacias (Institute for Space Research, Brazil)
IOC	Intergovernmental Oceanographic Commission (UNESCO)
IOP	Intensive Observation Period (HAPEX-Sahel)
IPCC	Intergovernmental Panel on Climate Change
IRMB	Institut Royal Météorologique de Belgique
ISA	Intermediate-Scale Area (GEWEX)
ISLSCP	International Satellite Land Surface Climatology Project
ITCZ	Intertropical Convergence Zone
IUGG	International Union of Geodesy and Geophysics
LAM	Local Area Model
LAMBADA	Large-scale Atmospheric Moisture Balance of Amazonia using Data Assimilation (LBA, Brazil)
LBA	Large-scale Biosphere–Atmosphere Experiments in Amazonia (consists of the experiments LAMBADA, BATERISTA and AMBIACE)
LMD	Laboratoire du Météorologie Dynamique (Paris, France)
LSA	Large-scale area (GEWEX)
LTER	Long-Term Ecological Research project (FIFE)
MAGS	Mackenzie GEWEX Study
MEDALUS	Mediterranean Desertification and Land Use project (EC)
MPI	Max Planck Institute for Meteorology (Hamburg, Germany)
MRI	Meteorological Research Institute (Japan)
NASA	National Aeronautics and Space Administration (USA)
NCAR	National Center for Atmospheric Research (Boulder, USA)
NDBA	Nested drainage basin approach
NEXRAD	Next generation of weather radar
NOAA	National Oceanographic and Atmospheric Administration (USA)
NOPEX	Nordic Hydrological Meteorological Experiment (Northern Hemisphere Climate Processes Land Surface Experiment)
NWSRFS	National Weather Service hydrologic model (USA)
OGCM	Ocean General Circulation Model

ORSTOM	Institut Français de Recherche Scientifique pour le Developpement et Coopération
OSU	Oregon State University
PAGES	PAst Global Changes (IGBP project)
PANASH	PAleoclimates of the Northern and Southern Hemispheres programme
PBMR	L-band and Push-Broom Microwave Radiometer
PEP3	Pole–Equator–Pole project (EU)
PET	Potential EvapoTranspiration
PILPS	Project for the Intercomparison of Land Surface Parameterization Schemes
PNA	Pacific/North America index
PNL	Pacific Northwest Laboratory (Seattle, Washington)
POLDER	POLarization and Directional profiling Radiometer)
PORTOS	A five-channel microwave profiling radiometer
RCM	Regional Circulation Model
REA	Representative Elementary Area
REKLIP	REgio-KLima-Projekt
RIVM	National Institute for Public Health and Environmental Protection (The Netherlands)
SACZ	South American Convergence Zone
SALSA-MEX	Semi-Arid Land Surface-Atmosphere Mountain Experiment
SHE	Système Hydrologique Européen model
SHI	State Hydrological Institute (Russia)
SOI	Southern Oscillation Index
SSA	Small-Scale Area (GEWEX)
SST	Sea Surface Temperature
SVAT	Soil, Vegetation and ATmosphere project (part of BAHC)
TDR	Time Domain Reflectometry
TOGA	Tropical Oceans Global Atmosphere Programme
TRMM	Tropical Rainfall Measurement Mission (satellite)
UFRGS	Federal University of Rio Grande do Sul (Brazil)
UKMO	United Kingdom Meteorological Office (Hadley Centre, Bracknell, UK)
UNEP	United Nations Environment Programme
UNESCO	United Nations Educational, Scientific and Cultural Organization
WCRP	World Climate Research Programme (joint undertaking of ICSU, WMO and IOC)
WCRP–Water	World Climate Research Programme–Water
WG-I	IPCC Working Group I on climate change
WG-II	IPCC Working Group II on the environmental and socio-economic impacts of climate change
WG-III	IPCC Working Group III on response strategies
WHYCOS	World HYdrological Cycle Observing System
WMO	World Meteorological Organization